ALASKA
TREES &
SHRUBS

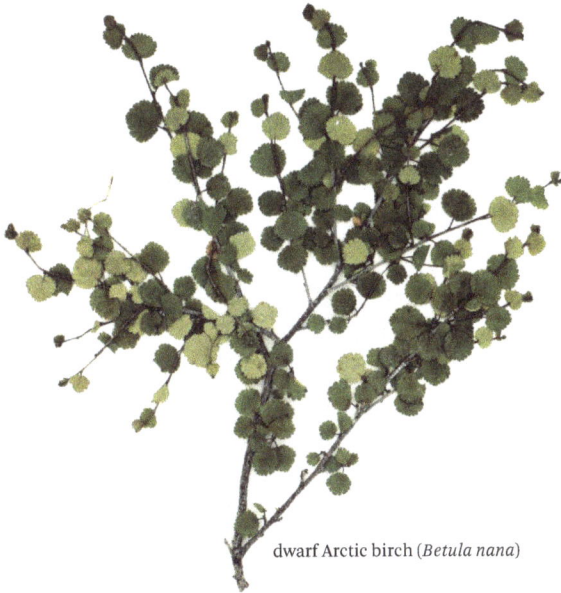

dwarf Arctic birch (*Betula nana*)

A FIELD GUIDE TO THE
WOODY PLANTS OF ALASKA

STEVE CHADDE

thinleaf alder (*Alnus incana*)

ALASKA TREES & SHRUBS
A Field Guide to the Woody Plants of Alaska

Steve W. Chadde

AN ORCHARD INNOVATIONS FIELD GUIDE
ISBN 978-1-951682-17-0

The author can be reached at: steve@chadde.net

VERSION 1.0, JANUARY 2020

CONTENTS

INTRODUCTION

A*laska Trees and Shrubs* describes more than 150 species of woody plants – the trees and shrubs – found in Alaska. Included are essentially all the native trees and shrubs of the state. Cultivated plants introduced from other areas, such as ornamental shrubs and fruits, are not included, apart from several introduced woody species which have escaped from cultivation and become established in the wild. Geographically, all of Alaska is included, from the narrow southeastern coastal region along the Pacific Ocean, west and southwest through the long chain of the Aleutian Islands, and north through the interior to the Arctic Ocean; also the many islands along the coasts. This reference will also be useful in northwestern Canada, including the Yukon and Northwest Territories, and northwestern British Columbia.

The Alaskan tree and shrub flora is represented by 19 plant families and 56 genera; the largest family being the willow family (Salicaceae), with 40 species (mostly willows, 37 species); with plants ranging in size from low, creeping, tundra shrubs, to taller shrubs and trees in interior and southeastern Alaska. Also well-represented are members of the heath family (Ericaceae), with 15 genera and 33 species; these include a number of blueberries (*Vaccinium*), bog plants such as species of *Andromeda, Kalmia,* and *Rhododendron,* and tundra and alpine species such as *Cassiope* and *Phyllodoce.*

This handbook was prepared for anyone desiring to learn more about Alaska's trees and shrubs and their habitats and uses. The intended audience ranges from those with little or no botanical background, to outdoors workers needing a comprehensive, up-to-date field reference. While technical terms have been kept to a minimum, a glossary is provided, and keys are included for all species, as well as keys for trees and shrubs in both summer and winter conditions. By using a combination of the keys, range maps, illustrations, descriptions, and habitat information, you should be able to confidently identify nearly every tree or shrub growing in the wild in Alaska. The most difficult group is the willows (*Salix*), and two keys are provided for the genus, one based on features of leaves and catkins, the other based on leaves, twigs, and growth form.

Trees are defined as woody plants having one erect perennial stem or trunk at least 3 in. (7.5 cm) in diameter at breast height (4½ feet, 1.4 m), a more or less definitely formed crown of foliage, and a height of at least 12 feet (4 m). Shrubs are woody plants smaller than trees, commonly with several perennial stems from the base. Among these are large or high shrubs and small or low shrubs. Also included are dwarf shrubs and subshrubs, creeping or prostrate plants with erect woody stems or woody only at their base, even if only 1–2 in. (2.5–5 cm.) above the ground (as in a number of tundra species). Woody vines, or plants with climbing stems supported on other plants, are not native to Alaska; the closest example is trailing black currant (*Ribes laxiflorum*), which has branches running along the ground and sometimes vinelike and climbing on shrubs. The state also has no woody plants poisonous to the touch or in contact with the skin. However, as noted, several species (also some herbs) have fruits or foliage poisonous if eaten. The only native species of wood epiphyte or parasite, western hemlock dwarf-mistletoe (*Arceuthobium tsugense*), is present in southeast Alaska.

Vast areas of Alaska have no native trees (e.g., vegetation types 5–10 on the map below). The tundra vegetation beyond the tree-line has a climate so severe that trees are absent. In many ways it corresponds to the treeless alpine zone on the high mountains. Northward, the number of shrub species also becomes less, and many species take the form of prostrate, creeping, or mat-forming shrubs, sometimes only several inches high.

A conspicuous feature of the Alaskan tree flora are the conifers (gymnosperms), which include the pines, spruces, firs, cedars, etc.; and 14 species are present in Alaska. Of these, all except one (tamarack) are evergreen with leaves reduced to needles or scales. Some species of creeping shrubs and low shrubs, particularly in the heath family, also have persistent leaves; these small plants may be covered by winter snows.

Of special interest are a few east Asian species reaching North America only in western Alaska. Two known in Alaska only on the westernmost Aleutian Islands are Siberian mountain-ash (*Sorbus sambucifolia*) of Siberia, and Miquel wintergreen (*Gaultheria pyroloides*) of Japan and adjacent Asia. Wedgeleaf willow (*Salix sphenophylla*) of Siberia is reported from the Seward Peninsula. Another Asiatic species common in the Aleutian Islands, and local eastward and northward in Alaska, is Kamchatka rhododendron (*Therorhodion camtschaticum*).

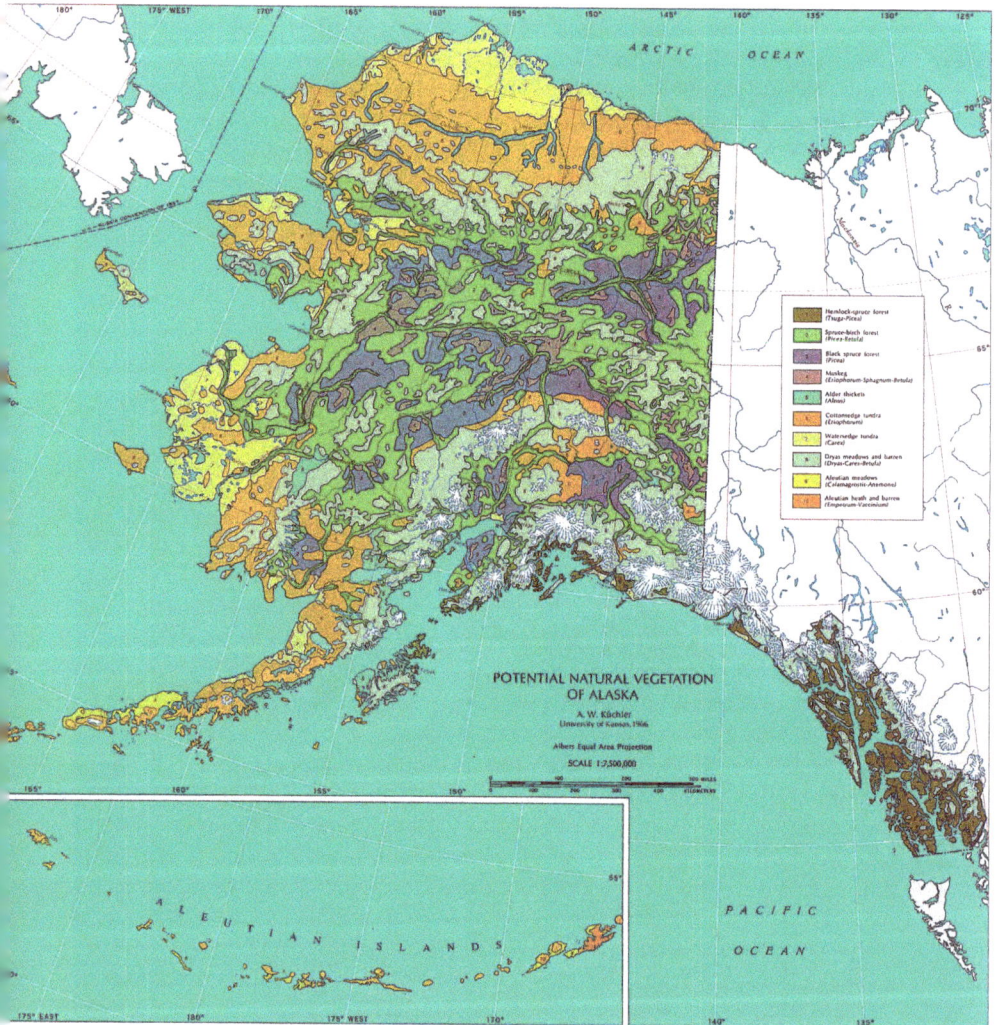

Potential natural vegetation of Alaska (Kuchler 1964). Kuchler defined potential natural vegetation as "the vegetation that would exist today if man were removed from the scene and if the plant succession after his removal were telescoped into a single moment."

VEGETATION OF ALASKA

Alaska is a land of contrasts contrasts in climate, physical geography, and vegetation. Containing 365.5 million acres (146 million hectares), Alaska has the highest mountain in North America as well as hundreds of square miles of boggy lowlands. The climate varies from mild and wet to cold and dry. Temperatures in the interior may range over 150° F. (83° C.) in a single year, and precipitation may be less than 10 inches (250 mm) annually. Conversely, southeastern coastal areas may receive 150 inches (3,800 mm) annual precipitation and have a temperature range of 70° F. (38° C.). Spanning nearly 1,300 miles (2,100 km) of latitude and 2,200 miles (3,500 km) of longitude, Alaska's vegetation varies from the towering, fast-growing forests of the southeastern coast, northward to the low, slow-growing boreal forests of the interior, to the treeless tundra of the north and west.

Of Alaska's land surface, approximately 119 million acres (48 million hectares) are forested. Of these, 28 million acres (11.2 million hectares) are classified as "commercial forests." These timber reserves provide the basis for one of the state's largest industries. At present, most of the state's timber production is from the Tongass and Chugach National Forests, which contain 92 percent of the commercial forests of coastal Alaska. Nearly all of the rest is from other areas within the coastal forests.

In addition to timber values, there are many other important uses of Alaska's forest and tundra areas. Much of Alaska is still wilderness, and the value of undisturbed wild areas may someday outweigh the potential value for producing lumber and pulp. Tourism and recreation in Alaska are important industries, based primarily on the state's scenic, wilderness, and wildlife values.

One of the most important resources of Alaska's forests are the wildlife species that inhabit them. The Alaska forests provide a home for large numbers of birds and mammals, most of which are dependent upon the woody plants, either directly or indirectly for food and shelter. Even those big game animals, such as the mountain sheep, mountain goat, and muskox, that spend much of their lives above or beyond tree-line, often use low woody plants for food during some part of the season.

Of the forest species, the moose is the most abundant and widespread large mammal of the interior forests; occasionally its range extends into the coastal areas. The moose survives the winter primarily by browsing on willows and other shrubs, especially in areas where the shrubs are growing thickly following forest fires, and in willow thickets along rivers. In coastal areas, the black-tail deer feeds primarily on blueberry and other shrubs during periods when the snow covers the lower vegetation. In the summer, the deer feeds mainly on the herbaceous plants that grow in the openings in the coastal forests. Even the caribou, often considered a tundra animal, spends the winters in the open forested area adjacent to tree-line, especially where lichen growth is abundant. In the summer, the caribou may browse several woody shrubs, especially resin and dwarf arctic birch and willows, as well as the herbaceous tundra vegetation. The small red squirrel, which is itself a source of food for larger furbearers, is dependent throughout the winter on seeds from spruce cones stored beneath the ground.

Several bird species survive through the Alaskan winters primarily by using woody plants as a source of food. Ptarmigan feed on willow and shrub birch buds, while ruffed and sharp-tailed grouse forage for berries from the past summer and feed on the buds of shrubs and trees. The spruce grouse of the interior and the blue grouse of the coastal areas live largely from the needles and buds of the spruce trees, as well as berries and buds of many shrubby species.

In the summer, insect life abounds in the forests and produces food for large numbers of small birds that nest and rear their young before migrating southward in the fall. In addition, the Alaskan forests and tundra are dotted with numerous lakes that serve as nesting places where large numbers of waterfowl rear their young during the short summer season.

The main vegetation types of Alaska and their most important trees and shrubs are listed below. A much more detailed classification of Alaska's vegetation is provided by Viereck et al. (1992).

■ Coastal Forests

The dense forests of western hemlock and Sitka spruce, a continuation of similar forests along the coast of British Columbia, Washington, and Oregon, extend about 900 miles (1,440 km) along the Alaska coast from the southeastern tip to Cook Inlet and Kodiak Island. Commercial stands occur from sea level to about 1,500 feet (460 m) elevation, but scattered trees rise to a timberline at 2,000 to 3,000 feet (460–915 m).

Coastal forests are characterized by steep, rough topography. In many areas only a narrow band of trees exists between the ocean and the tundra on snowclad mountains above. The scenic grandeur of the region is unsurpassed. The narrow waterways with steep forested slopes, the rugged high mountains, and the many glaciers reaching to the coast through forested valleys along with an abundance of streams and lakes offer a wealth of recreation values to Alaskans and tourists.

The climate is cool and cloudy in summer, and the winters are mild. Snowfall may be heavy in some forested areas in the northern part, but much of the high precipitation falls as rain. Annual precipitation varies from as much as 222 inches (5,640 mm) on the outside coast of the southeasternmost islands to 25 inches (630 mm) at Homer on the boundary between coastal and interior forests. The mean annual temperature in the coastal forests ranges from 46° F. (8° C.) at Ketchikan to 37° F. (3° C.) at Cordova. Summer temperatures range in the upper 50's (13–16° C.) and winter temperatures for the coldest month range from the low 20's (-6° C.) to the mid 30's (+2° C.).

Coastal Spruce-Hemlock Forests

In the southern part, the coastal forests are composed primarily of **western hemlock and Sitka spruce** with a scattering of mountain hemlock, western redcedar, and Alaska-cedar. Red alder is common along streams, beach fringes, and on soils recently disturbed by logging and landslides. Black cottonwood grows on the flood plains of major rivers and recently deglaciated areas on the mainland. Subalpine fir and Pacific silver fir occur occasionally at tree line and near sea level but are not abundant enough to be of commercial value. Blueberries, huckleberry, copperbush, devil's-club, and salal are the most important shrubs. Because of the high rainfall and resulting high humidity, mosses grow in great profusion on the ground, on fallen logs, and on the lower branches of trees, as well as in forest openings.

In poorly drained areas at low elevations, open muskegs of low shrubs, sedges, grasses, and mosses are common. These areas are treeless or may have a few scattered shrubby trees of shore pine (lodgepole pine), western hemlock, mountain hemlock, Alaska-cedar, and Sitka spruce.

In the northern and western sections of the coastal forests, the makeup of the tree species changes. Western redcedar is not found north of Frederick Sound, and Alaska-cedar drops out at Prince William Sound. Cottonwood is extensive along some of the glacial outwash rivers and becomes commercially important in the Haines area and on the alluvial terraces to the west. Western hemlock becomes of less importance westward but is found as far as Cook Inlet. Only Sitka spruce remains as the important tree in the coastal forests west of

Cook Inlet and the lone conifer on Afognak and Kodiak Islands. Douglas-fir, which is characteristic of the coastal forests of Oregon, Washington, and southern British Columbia, does not reach Alaska. Common trees and shrubs of the coastal forests are:

Trees	stink currant *Ribes bracteosum*
red alder *Alnus rubra*	trailing black currant *Ribes laxiflorum*
Alaska-cedar *Callitropis nootkatensis*	western thimbleberry *Rubus parviflorus*
Sitka spruce *Picea sitchensis*	salmonberry *Rubus spectabilis*
black cottonwood *Populus trichocarpa*	Barclay willow *Salix barclayi*
western hemlock *Tsuga heterophylla*	Scouler willow *Salix scouleriana*
mountain hemlock *Tsuga mertensiana*	Sitka willow *Salix sitchensis*
	Pacific red elder *Sambucus racemosa*
Shrubs	Alaska blueberry *Vaccinium alaskaense*
Sitka alder *Alnus viridis*	dwarf blueberry *Vaccinium cespitosum*
salal *Gaultheria shallon*	early blueberry *Vaccinium ovalifolium*
rusty menziesia *Menziesia ferruginea*	red huckleberry *Vaccinium parvifolium*
devil's-club *Oplopanax horridus*	squashberry *Viburnum edule*

▓ Interior Forests

The **white spruce-paper birch** forest, extending from the Kenai Peninsula to the south slopes of the Brooks Range and westward nearly to the Bering Sea, is called the boreal forest or taiga the Russian equivalent. These forests cover about 32 percent of the area or about 106 million acres (42.4 million hectares). However, only about one-fifth or 22.5 million acres (9 million hectares) are classified as commercial forest land.

Characteristic forest stands are found in the Tanana and Yukon Valleys. Here, in contrast to the coast, climatic conditions are extreme. The mean annual temperature is 20 to 30° F. (-7° C. to -1° C.) but winter temperatures below -40° F. (-40° C.) are common and the coldest month averages -10 to -20° F. (-23° to -29° C.). In contrast, summer temperatures may reach into the 90's (above 30° C.), and the warmest month of the year has an average of 60° F. (16° C.). Permanently frozen ground is of scattered occurrence in the southern part of the interior forests and nearly continuous in the northern sections. Although precipitation is light, 6 to 12 inches (150–300 mm) per year, evaporation is low and permafrost forms an impervious layer so that bogs and wet areas are common. Snowfall averages 55 inches (140 cm) per year at Fairbanks, but snow cover usually persists from mid-October until mid- to late-April. Day-length is also extreme in the boreal forest regions with nearly 24 hours of daylight available for plant growth in June, but with only a few hours of sunlight during the winter months. Forest fires have always been an important aspect of the environment of the Alaska interior forests. Even now with modern fire detecting and fighting techniques, more than 4 million acres may burn in a single dry summer.

The vegetation types in interior Alaska form a mosiac of patterns that is related in part to past fire history, to slope and aspect, and to the presence or absence of permafrost. Most forest stands are mixtures of two or more tree species but are usually classified by the dominant species.

Closed Spruce-Hardwood Forests

White spruce type

In general, the best commercial stands of **white spruce** are found on the warm, dry, south-facing hillsides and adjacent to rivers where drainage is good and permafrost lacking. These stands are rather open under the canopy but may contain shrubs of rose, alder, and willow. The forest floor is usually carpeted with a thick moss mat. On the better sites, 100 to 200 year-old spruce with diameters of 10–24 inches (25–60 cm) may average 10,000 board feet per acre

(58 cubic meters per hectare). Stands in which commercial white spruce are dominant occupy 12.8 million acres (5.1 million hectares) in interior Alaska. The most common trees and shrubs of the white spruce type are:

Trees	mountain-cranberry *Vaccinium vitis-idaea*
paper birch *Betula papyrifera*	bog blueberry *Vaccinium uliginosum*
white spruce *Picea glauca*	squashberry *Viburnum edule*
balsam poplar *Populus balsamifera*	
	Occasional to rare shrubs
Common shrubs	bearberry *Arctostaphylos uva-ursi*
red-fruit bearberry *Arctous rubra*	resin birch *Betula glandulosa*
crowberry *Empetrum nigrum*	bush cinquefoil *Dasiphora fruticosa*
narrow-leaf Labrador-tea *Rhododendron*	rusty menziesia *Menziesia ferruginea*
tomentosum	tall blueberry willow *Salix boothii*
American red currant *Ribes triste*	grayleaf willow *Salix glauca*
prickly rose *Rosa acicularis*	halberd willow *Salix hastata*
feltleaf willow *Salix alaxensis*	Richardson's willow *Salix richardsonii*
littletree willow *Salix arbusculoides*	park willow *Salix monticola*
Bebb willow *Salix bebbiana*	Scouler willow *Salix scouleriana*
buffaloberry *Shepherdia canadensis*	dwarf blueberry *Vaccinium caespitosum*

Recent burns

Because of extensive burns during the past 100 years, large areas of the interior are in various stages of forest succession. The succession that follows fire is varied and depends upon topography, previous vegetation, severity of burn, and available seed source at the time of burn. In general, fires are followed by a shrubby stage consisting primarily of light-seeded willows. The most important woody plants to follow immediately after fire are:

Common shrubs	**Occasional to rare shrubs**
Labrador-tea *Rhododendron groenlandicum*	thinleaf alder *Alnus incana*
narrow-leaf Labrador-tea *Rhododendron*	Sitka alder *Alnus viridis*
tomentosum	red-fruit bearberry *Arctous rubra*
prickly rose *Rosa acicularis*	bearberry *Arctostaphylos uva-ursi*
littletree willow *Salix arbusculoides*	bush cinquefoil *Dasiphora fruticosa*
Barclay willow *Salix barclayi*	crowberry *Empetrum nigrum*
Bebb willow *Salix bebbiana*	American red currant *Ribes triste*
Scouler willow *Salix scouleriana*	buffaloberry *Shepherdia canadensis*
dwarf blueberry *Vaccinium cespitosum*	Beauverd spirea *Spiraea stevenii*
mountain-cranberry *Vaccinium vitis-idaea*	bog blueberry *Vaccinium uliginosum*

Quaking aspen type

Following fire and a willow stage, fast-growing **quaking aspen** stands develop in upland areas on south facing slopes. The aspen mature in 60 to 80 years and are eventually replaced by white spruce, except in excessively dry sites where they may persist. Occasionally aspen stands also follow fire on well drained lowland river terraces and, in these situations, are usually replaced by black spruce in the successional sequence. Stands with aspen dominant occupy about 2.4 million acres (960,000 hectares) in central Alaska. Woody plants occurring in the aspen type are:

Trees	Occasional to rare shrubs
white spruce *Picea glauca*	red-fruit bearberry *Arctous rubra*
black spruce *Picea mariana*	Alaska sagebrush *Artemisia alaskana*
quaking aspen *Populus tremuloides*	fringed sagebrush *Artemisia frigida*
	resin birch *Betula glandulosa*
Common shrubs	crowberry *Empetrum nigrum*
bearberry *Arctostaphylos uva-ursi*	common juniper *Juniperus communis*
prickly rose *Rosa acicularis*	Labrador-tea *Rhododendron groenlandicum*
Bebb willow *Salix bebbiana*	American red raspberry *Rubus idaeus*
Scouler willow *Salix scouleriana*	dwarf blueberry *Vaccinium cespitosum*
buffaloberry *Shepherdia canadensis*	bog blueberry *Vaccinium uliginosum*
mountain-cranberry *Vaccinium vitis-idaea*	squashberry *Viburnum edule*

Paper birch type

Paper birch is the common invading tree after fire on east- and west-facing slopes and occasionally on north slopes and flat areas. This species occurs either in pure stands or more often is mixed with white spruce, aspen, or black spruce. Shrubs may be similar to those under aspen but usually Labrador-tea and mountain-cranberry are more common. Paper birch may be 60–80 feet (18–24 m) tall and have diameters up to 18 inches (46 cm), but an average diameter of 8–9 inches (20–22 cm) is more common in the interior birch stands. Stands dominated by paper birch occupy about 5 million acres (2 million hectares) of interior forests and are especially widespread in the Susitna River Valley. Trees and shrubs occurring in the birch type are:

Trees	Scouler willow *Salix scouleriana*
paper birch *Betula papyrifera*	mountain-cranberry *Vaccinium vitis-idaea*
white spruce *Picea glauca*	dwarf blueberry *Vaccinium cespitosum*
black spruce *Picea mariana*	squashberry *Viburnum edule*
Common shrubs	**Occasional to rare shrubs**
Labrador-tea *Rhododendron groenlandicum*	crowberry *Empetrum nigrum*
narrow-leaf Labrador-tea *Rhododendron tomentosum*	rusty menziesia *Menziesia ferruginea*
American red currant *Ribes triste*	devil's-club *Oplopanax horridus*
prickly rose *Rosa acicularis*	northern black currant *Ribes hudsonianum*
Barclay willow *Salix barclayi*	American red raspberry *Rubus idaeus*
Bebb willow *Salix bebbiana*	Pacific red elder *Sambucus racemosa*
	Cascade mountain-ash *Sorbus scopulina*

Balsam poplar type

Another tree species of importance within the closed spruce-hardwood forest in interior Alaska is **balsam poplar**, which reaches its greatest size and abundance on the floodplains of meandering glacial rivers. It invades sandbars and grows rapidly to heights of 80–100 feet (24–30 m) and diameters of 24 inches (60 cm) before being replaced by white spruce. Balsam poplar also occurs in small clumps near the altitudinal and latitudinal limit of trees in the Alaska Range and north of the Brooks Range. Commercial stands occupy 2.1 million acres (840,000 hectares), primarily along the Yukon, Tanana, Susitna, and Kuskokwim Rivers. In the Susitna Valley balsam poplar is often replaced in this type by black cottonwood or by hybrids between the two. Woody plants of this type include:

Trees	prickly rose *Rosa acicularis*
balsam poplar *Populus balsamifera*	squashberry *Viburnum edule*
black cottonwood *Populus trichocarpa*	
white spruce *Picea glauca*	**Occasional to rare shrubs**
	bush cinquefoil *Dasiphora fruticosa*
Common shrubs	silverberry *Eleagnus commutata*
Sitka alder *Alnus viridis*	high blueberry willow *Salix boothii*
thinleaf alder *Alnus tenuifolia*	Scouler willow *Salix scouleriana*
littletree willow *Salix arbusculoides*	buffaloberry *Shepherdia canadensis*
feltleaf willow *Salix alaxensis*	

Open, Low-Growing Spruce Forests

On north-facing slopes and poorly drained lowlands, forest succession leads to boggy, open black spruce stands, usually underlain by permafrost. The black spruce are slow-growing and seldom exceed 8 inches (20 cm) in diameter (and are usually much smaller); a tree 2 inches (5 cm) in diameter is often 100 years old. Black spruce comes in abundantly after fire because its persistent cones open after a fire and spread abundant seed over the burned areas. A thick moss mat, often of sphagnum mosses, sedges, grasses, and heath or ericaceous shrubs usually make up the subordinate vegetation of open black spruce stands. Associated with black spruce in the wet bottomlands is the slow-growing tamarack. As with black spruce, it is of little commercial value, seldom reaching a diameter of more than 6 inches (15 cm). Woody plants of these low-growing spruce forests include:

Trees	diamondleaf willow *Salix pulchra*
black spruce *Picea mariana*	Scouler willow *Salix scouleriana*
tamarack *Larix laricina*	bog blueberry *Vaccinium uliginosum*
paper birch *Betula papyrifera*	mountain-cranberry *Vaccinium vitis-idaea*
white spruce *Picea glauca*	
	Occasional to rare shrubs
Common shrubs	resin birch *Betula glandulosa*
red-fruit bearberry *Arctous rubra*	dwarf arctic birch *Betula nana*
crowberry *Empetrum nigrum*	bush cinquefoil *Dasiphora fruticosa*
Labrador-tea *Rhododendron groenlandicum*	rusty menziesia *Menziesia ferruginea*
prickly rose *Rosa acicularis*	narrow-leaf Labrador-tea *Rhododendron*
littletree willow *Salix arbusculoides*	*tomentosum*
Bebb willow *Salix bebbiana*	dwarf blueberry *Vaccinium cespitosum*
grayleaf willow *Salix glauca*	bog cranberry *Vaccinium oxycoccos*
low blueberry willow *Salix myrtillifolia*	

Treeless Bogs

Coastal areas

Within the coastal forests in depressions, flat areas, and on some gentle slopes where drainage is poor, treeless areas occur. The vegetation is variable but most commonly consists of a thick sphagnum moss mat, sedges, rushes, low shrubs, and fruticose lichens. This type is locally called "muskeg." Often a few slow-growing, poorly formed, shore pine, western hemlock, or Alaska-cedar may be present where conditions are drier. In more exposed situations and in the driest areas, shrubs may be dominant over the sedge and herbaceous mat. Ponds are often present in the peaty substrate. Characteristic shrubs of coastal Alaska bogs include:

bog-rosemary *Andromeda polifolia*	Barclay willow *Salix barclayi*
crowberry *Empetrum nigrum*	undergreen willow *Salix commutata*
common juniper *Juniperus communis*	bog cranberry *Vaccinium oxycoccos*
bog kalmia *Kalmia polifolia*	bog blueberry *Vaccinium uliginosum*
rusty menziesia *Menziesia ferruginea*	mountain-cranberry *Vaccinium vitis-idaea*
Labrador-tea *Rhododendron groenlandicum*	

Interior areas

Within the boreal forest are extensive bogs where conditions are too wet for tree growth. North of the Alaska Range in unglaciated areas, bogs occur on old river terraces and outwash, in-filling ponds and old sloughs, and occasionally on gentle, north-facing slopes. Bogs are common south of the Alaska Range, on the fine clay soils formed in former glacial lake basins, and on morainal soils within the glaciated area. Bogs are also common on the extensive flat lands of the lower Yukon and Kuskokwim Rivers.

The vegetation of these bogs consists of varying amounts of grasses, sedges, and mosses, especially sphagnum. Often the surface is made uneven by stringlike ridges. Much of the surface of these bogs is too wet for shrubs but on the drier peat ridges are a number of heath or ericaceous shrubs, willows, and dwarf birches. The woody plants of the treeless bogs include:

Common shrubs	bog cranberry *Vaccinium oxycoccos*
bog-rosemary *Andromeda polifolia*	bog blueberry *Vaccinium uliginosum*
resin birch *Betula glandulosa*	mountain-cranberry *Vaccinium vitis-idaea*
dwarf arctic birch *Betula nana*	
leatherleaf *Chamaedaphne calyculata*	**Occasional to rare shrubs**
sweetgale *Myrica gale*	thinleaf alder *Alnus incana*
Labrador-tea *Rhododendron groenlandicum*	Sitka alder *Alnus viridis*
narrow-leaf Labrador-tea *Rhododendron*	red-fruit bearberry *Arctous rubra*
tomentosum	bush cinquefoil *Dasiphora fruticosa*
Barclay willow *Salix barclayi*	crowberry *Empetrum nigrum*
Alaska bog willow *Salix fuscescens*	grayleaf willow *Salix glauca*
low blueberry willow *Salix myrtillifolia*	netleaf willow *Salix reticulata*
diamondleaf willow *Salix pulchra*	Beauverd spirea *Spiraea stevenii*

Shrub Thickets
Coastal alder thickets

Dense thickets of shrubs can be found within all major vegetation zones in Alaska. In coastal Alaska, there are extensive alder thickets between the beach and the forest, between tree-line and alpine tundra meadows, and extending from tree-line downward through the forest in avalanche tracks and along streams. The shrub thicket type is also common in southeastern Alaska in clearcut areas. Alder thickets are almost impenetrable as the boles of the shrubs tend to grow horizontally as well as vertically. To travel through the thicket is even worse; the spiny devil's-club and salmonberry are frequently present. Beneath the alder there is often a well-developed grass and fern layer, as well as a number of herbs and shrubs. The most common woody plants in this type are:

Sitka alder *Alnus viridis*	Barclay willow *Salix barclayi*
luetkea *Luetkea pectinata*	Scouler willow *Salix scouleriana*
Oregon crab apple *Malus fusca*	Sitka willow *Salix sitchensis*
rusty menziesia *Menziesia ferruginea*	Pacific red elder *Sambucus racemosa*
devil's-club *Oplopanax horridus*	Sitka mountain-ash *Sorbus sitchensis*
stink currant *Ribes bracteosum*	Alaska blueberry *Vaccinium alaskaense*
trailing black currant *Ribes laxiflorum*	dwarf blueberry *Vaccinium cespitosum*
Nootka rose *Rosa nutkana*	early blueberry *Vaccinium ovalifolium*
Western thimbleberry *Rubus parviflorus*	red huckleberry *Vaccinium parvifolium*
salmonberry *Rubus spectabilis*	

Floodplain thickets

Another major shrub type, **floodplain thickets**, is found along rivers. Although somewhat different in species composition, the type is rather similar from the rivers of the southern coastal areas to the broad braided rivers north of the Brooks Range. This type forms on newly exposed alluvial deposits that are periodically flooded. It develops quickly and may reach heights of 15 to 20 feet (4.5–6 m) in the south and central Alaska, and 5 to 10 feet (1.5–3 m) along the rivers north of the Brooks Range. The main dominant shrubs of this type are willows and occasionally alders, with a number of lower shrubs under the canopy. Shrubs of this type include:

thinleaf alder *Alnus incana*	grayleaf willow *Salix glauca*
Sitka alder *Alnus viridis*	halberd willow *Salix hastata*
red-osier dogwood *Cornus alba*	sandbar willow *Salix interior*
silverberry *Eleagnus commutata*	Richardson's willow *Salix richardsonii*
sweetgale *Myrica gale*	Pacific willow *Salix lasiandra*
prickly rose *Rosa acicularis*	park willow *Salix monticola*
American red raspberry *Rubus idaeus*	tall blueberry willow *Salix boothii*
feltleaf willow *Salix alaxensis*	diamondleaf willow *Salix pulchra*
littletree willow *Salix arbusculoides*	Setchell willow *Salix setchelliana*
Barclay willow *Salix barclayi*	Sitka willow *Salix sitchensis*
Bebb willow *Salix bebbiana*	buffaloberry *Shepherdia canadensis*
barren-ground willow *Salix niphoclada*	squashberry *Viburnum edule*
undergreen willow *Salix commutata*	

Birch-alder-willow thickets

A third type of shrub thicket occurs near tree line in interior Alaska and beyond tree line in extensive areas of the Alaska and Seward peninsulas. It consists of resin birch, alder, and several willow species, usually forming thickets 3 to 10 feet (1–3 m) tall. The thickets may be extremely dense, or they may be open and interspersed with reindeer lichens, low heath - type shrubs, or patches of alpine tundra. Alders tend to occupy the wetter sites, birch the mesic sites, with tundra vegetation in drier or wind-exposed areas. The type extends below tree-line where it is often associated with widely spaced white spruce. Shrubs of this type include:

Sitka alder *Alnus sinuata*	Barclay willow *Salix barclayi*
alpine bearberry *Arctous alpina*	undergreen willow *Salix commutata*
resin birch *Betula glandulosa*	Alaskan bog willow *Salix fuscescens*
dwarf arctic birch *Betula nana*	diamondleaf willow *Salix pulchra*
bush cinquefoil *Dasiphora fruticosa*	Richardson's willow *Salix richardsonii*
crowberry *Empetrum nigrum*	netleaf willow *Salix reticulata*
narrow-leaf Labrador-tea *Rhododendron*	Beauverd spirea *Spiraea stevenii*
tomentosum	bog blueberry *Vaccinium uliginosum*

■ Tundra

Low tundra vegetation can be divided into three main types: **moist tundra, wet tundra**, and **alpine tundra**. Within each of these major types are mosaics of subtypes related to differences in topography, slope, aspect, and substrate.

Moist Tundra

Moist tundra occupies the foothills and lower elevations of the Alaska Range as well as extensive areas on the Seward and Alaska peninsulas, the Aleutian Islands, and the islands of the Bering Sea. The type varies from almost continuous and uniformly developed cottongrass (*Eriophorum*) tussocks with sparse growth of other sedges and dwarf shrubs, to stands where tussocks are scarce or lacking and dwarf shrubs are dominant. Over wide areas in Arctic Alaska, the cottongrass tussock type is the most widespread of all vegetation types. In northern areas the type is often dissected by polygonal patterns created by underlying ice wedges. On the Aleutian Islands, the type consists of tall grass meadows interspersed with a dense low heath shrub type. The shrubs found in this type from the Aleutian Islands to the north slope of the Brooks Range include:

Sitka alder *Alnus viridis*	undergreen willow *Salix commutata*
alpine bearberry *Arctous alpina*	Alaska bog willow *Salix fuscescens*
resin birch *Betula glandulosa*	grayleaf willow *Salix glauca*
dwarf arctic birch *Betula nana*	Richardson's willow *Salix richardsonii*
four-angled cassiope *Cassiope tetragona*	diamondleaf willow *Salix pulchra*
entire-leaf mountain-avens *Dryas integrifolia*	ovalleaf willow *Salix ovalifolia*
white mountain-avens *Dryas octopetala*	polar willow *Salix polaris*
alpine-azalea *Kalmia procumbens*	netleaf willow *Salix reticulata*
Aleutian mountain-heath *Phyllodoce aleutica*	least willow *Salix rotundifolia*
Lapland rosebay *Rhododendron lapponicum*	Beuverd spirea *Spiraea stevenii*
narrow-leaf Labrador-tea *Rhododendron*	Kamchatka rhododendron *Therorhodion*
tomentosum	*camtschaticum*
arctic willow *Salix arctica*	bog cranberry *Vaccinium oxycoccos*
Barclay willow *Salix barclayi*	bog blueberry *Vaccinium uliginosum*
Barratt willow *Salix barrattiana*	mountain-cranberry *Vaccinium vitis-idaea*
Chamisso willow *Salix chamissonis*	

Wet Tundra

The **wet tundra** type also includes the low coastal marshes of southern Alaska. The type is most extensive along the coastal plain north of the Brooks Range, the northern part of the Seward Peninsula, and on the broad Yukon delta. It is usually found in areas with many shallow lakes and little topographic relief. Standing water is almost always present in the summer, and in the northern parts permafrost is close to the surface. Micro-relief is provided by peat ridges and polygonal features related to frost action and ice wedges. The vegetation is primarily a sedge and cottongrass mat, usually not formed into tussocks. The few woody plants occur on the driest sites where the microrelief raises them above the standing water table. Shrubs include:

bog-rosemary *Andromeda polifolia*	netleaf willow *Salix reticulata*
resin birch *Betula glandulosa*	Richardson's willow *Salix richardsonii*
dwarf arctic birch *Betula nana*	ovalleaf willow *Salix ovalifolium*
narrow-leaf Labrador-tea *Rhododendron tomentosum*	bog cranberry *Vaccinium oxycoccos*
Alaska bog willow *Salix fuscescens*	bog blueberry *Vaccinium uliginosum*
diamondleaf willow *Salix pulchra*	mountain-cranberry *Vaccinium vitis-idaea*

Alpine Tundra

In all mountain ranges of Alaska and on exposed ridges in the arctic and southwestern coastal areas, there is a zone of **alpine tundra**. Much of this type consists of barren rocks, but interspersed between the rocks and rubble are low mat plants, both herbaceous and shrubby. Dominant in this type in northern areas and in the Alaska Range are low mats of eight-petal mountain-avens, which may cover entire ridges and slopes, along with many mat forming herbs, such as moss-campion (*Silene acaulis*), black oxytrope (*Oxytropis bryophila*), arctic sandwort (*Minuartia arctica*), and several grasses and sedges. In the southeastern coastal mountains and the Aleutians, the most important plants are the low heath shrubs, especially cassiopes and mountain-heaths. They are most abundant where snow accumulates in the winter and lingers into late-spring. On the Aleutian Islands, this type consists primarily of crowberry, bog blueberry, mountain-cranberry, alpine-azalea, and several dwarf willows. Shrubs of alpine tundra throughout Alaska include:

alpine bearberry *Arctous alpina*	narrow-leaf Labrador-tea *Rhododendron tomentosum*
resin birch *Betula glandulosa*	arctic willow *Salix arctica*
dwarf arctic birch *Betula nana*	Chamisso willow *Salix chamissonis*
Alaska cassiope *Cassiope lycopodioides*	Alaska bog willow *Salix fuscescens*
Mertens cassiope *Cassiope mertensiana*	ovalleaf willow *Salix ovalifolia*
starry cassiope *Cassiope stelleriana*	skeletonleaf willow *Salix phlebophylla*
four-angled cassiope *Cassiope tetragona*	polar willow *Salix polaris*
diapensia *Diapensia lapponica*	diamondleaf willow *Salix pulchra*
eight-petal mountain-avens *Dryas octopetala*	netleaf willow *Salix reticulata*
entire-leaf mountain-avens *Dryas integrifolia*	least willow *Salix rotundifolia*
crowberry *Empetrum nigrum*	Kamchatka rhododendron *Therorhodion camtschaticum*
alpine-azalea *Kalmia procumbens*	dwarf blueberry *Vaccinium cespitosum*
luetkea *Luetkea pectinata*	bog blueberry *Vaccinium uliginosum*
Aleutian mountain-heath *Phyllodoce aleutica*	mountain-cranberry *Vaccinium vitis-idaea*
blue mountain-heath *Phyllodoce coerulea*	
red mountain-heath *Phyllodoce empetriformis*	
Lapland rosebay *Rhododendron lapponicum*	

NOTE On the following pages, the descriptions are arranged alphabetically, first by scientific name of the plant family (e.g., Pinaceae), then alphabetically by genus and species. Keys to all trees and shrubs begin on page 229.

ALASKA TREES & SHRUBS—GYMNOSPERMS

Seed plants with seeds partly exposed (gymnosperms), not enclosed in fruits, are represented in Alaska by three families of conifers or softwoods, the **yew family** (Taxaceae), the **pine family** (Pinaceae), and the **cypress family** (Cupressaceae). The Alaska examples are evergreen (with one exception) trees and shrubs with narrow or small leaves resembling needles or scales. **Pacific yew** (*Taxus brevifolia*), the Alaska member of the yew family, is distinguished by the brown seeds borne singly in a scarlet, juicy, cuplike or berrylike disk, by the flat, pointed, nonresinous needles in 2 rows, and by the twisted leafstalks extending down the twig.

■ CYPRESS FAMILY *Cupressaceae*

The cypress family (Cupressaceae) has two genera and species of trees in Alaska, also a third genus with two species of low shrubs, the junipers. This family, formerly included in the pine family, is characterized by small scalelike leaves paired or in 3's. The **cones** are small with few cone-scales bearing mostly few seeds with short side wings. However, junipers have berrylike cones and wingless seeds. Characteristics of our three genera are:

  • **Western redcedar** (*Thuja plicata*), the only Alaska species of *Thuja*. Leaves scalelike, flattened and curved, on flattened twigs in fanlike sprays. Small cones clustered near ends of twigs and becoming turned up.

  • **Alaska-cedar** (*Callitropsis nootkatensis*), the only species of white-cedar. Leaves scalelike, pointed and spreading. Leafy twigs 4-angled or slightly flattened. Cones small, hard, nearly round.

  • **Juniper** (*Juniperus*), 2 species, common juniper and creeping juniper. Dwarf shrubs with scalelike or awl-shaped leaves, small round berrylike cones, and few wingless seeds.

Alaska-Cedar
Callitropsis nootkatensis (D. Don) D.P. Little

OTHER NAMES Alaska yellow-cedar, Nootka false-cypress, yellow-cedar, Alaska cypress, Sitka cypress, yellow cypress
SYNONYMS *Chamaecyparis nootkatensis* (D. Don) Spach, *Cupressus nootkatensis* D. Don, *Xanthocyparis nootkatensis* (D. Don) Farjon & D.K. Harder
DESCRIPTION Medium sized evergreen tree 40–80 feet (12–24 m) high and 1–2 feet (30–60 cm) in trunk diameter, sometimes a large tree to 100 feet (30 m) tall and 4 feet (1.2 m) in diameter, with narrow crown of slightly drooping branches.

LEAVES scalelike, 1/16–1/8 in. (1.5–3 mm) long, pointed and spreading, yellow-green, with slightly spreading, pointed tips; leaves on sharp-pointed leader twigs to 1/4 in. (6 mm) long.
LEAFY TWIGS 4-angled or slightly flattened; in flat, spreading sprays on drooping slightly branches, becoming reddish brown.
BARK shreddy, with long narrow shreds and fissures, ash gray or purplish brown.
WOOD with distinctive odor, relatively straight-grained, moderately heavy, moderately hard. Heartwood bright yellow with narrow band of lighter sapwood.

ALASKA-CEDAR

CONES scattered, short-stalked, nearly round, less than 1/2 in. (12 mm) in diameter, hard, ashy gray, often covered with whitish bloom, of 4 or 6 paired rounded hard cone-scales each with a central pointed projection, maturing in 2 years. **Seeds** 2–4 under a cone-scale, 3/16 in. (4 mm) long, brown, with 2 broad wings.

HABITAT Alaska-cedar extends along the coast of southeast Alaska from sea level to timberline but is best developed at 500–1,200 feet altitude. It is scattered with western redcedar, in pure stands, in forests of Sitka spruce and western hemlock, and, on higher slopes or muskegs, with mountain hemlock. The trees are slow-growing, those 15–20 in. (38–51 cm) in trunk diameter being 200–300 years old.

USES The very durable aromatic wood is easily worked and takes a beautiful finish. It is valuable for window frames and exterior doors, boat construction, and similar purposes. It is used also for utility poles, piles, interior finish, furniture, cabinet work, patterns, and novelties. Alaska Natives of southeast Alaska made their canoe paddles from this wood. Much Alaska-cedar is exported to Japan in log form, though some is used locally.

ALASKA-CEDAR

ALASKA-CEDAR - CONE

ALASKA-CEDAR BARK

JUNIPER *Juniperus*

Low or prostrate aromatic evergreen shrubs (elsewhere also trees). **Leaves** opposite in 4 rows or in 3's, crowded, scalelike, blunt, and closely pressed against twig or awl-shaped, sharp-pointed, and spreading. Male and female cones mostly on different plants. **Cones** small, berrylike, fleshy, round, 1/4–3/8 in. (6–10 mm) in diameter, mostly blue, fleshy, resinous, not opening, containing usually 1–4 wingless seeds.

KEY TO ALASKA JUNIPERS

1   Leaves awl-shaped, sharp-pointed, spreading, in groups of 3
                 COMMON JUNIPER (*Juniperus communis*)
1   Leaves mostly scalelike, blunt, pressed against twig, paired
                 CREEPING JUNIPER (*Juniperus horizontalis*)

Common Juniper
Juniperus communis L.

OTHER NAMES low juniper, mountain common juniper

DESCRIPTION Low or prostrate spreading evergreen shrub to 2 feet (0.6 m) high, forming mats or clumps to 10 feet (3 m) in diameter.

LEAVES in groups of 3 (whorled), spreading at right angles or curved slightly downward, awl-shaped, 1/4–1/2 in. (6–12 mm) long, less than 1/16 in. (1.5 mm) wide, stiff, very sharp-pointed, jointed at base, whitish and grooved above, shiny yellow-green beneath.

TWIGS slender, 3–angled, light yellow, hairless.

BARK gray or dark reddish brown, rough, scaly and thin.

CONES lateral on very short scaly stalks, berry-like, round, 1/4–3/8 in. (6–10 mm) in diameter, blue and covered with a bloom, hard, mealy, resinous and sweetish, maturing in 2 or 3 years and persistent. **Seeds** 3 or fewer, light brown, more than 1/8 in. (3 mm) long, pointed.

HABITAT Scattered to rare in rocky tundra, sunny slopes, sandy areas, and forest openings.

NOTE Including a few geographic varieties, this species is the most widely distributed conifer in the world, and the most widespread tree species in the north temperate zone. Common juniper becomes a small tree rarely in New England and frequently in Europe.

USES Juniper is planted as an ornamental in Alaska, mostly as a ground cover in dry and rocky locations. In northern Europe the fruits have been used to flavor gin.

COMMON JUNIPER

Creeping Juniper
Juniperus horizontalis Moench

OTHER NAME creeping savin

DESCRIPTION Prostrate or trailing evergreen shrub with long horizontal stems often rooting and with short erect twigs 2–6 in. (5–15 cm) high.

LEAVES paired in 4 rows, mostly scalelike, 1/16 in. (1.5 mm) long, blunt and short-pointed with gland dot, blue green, shedding with twigs, on young plants and leaders awl-shaped, sharp-pointed, 3/16–1/4 in. long (5–6 mm) long.

TWIGS less than 1/16 in. (1.5 mm) broad, 4-angled, covered with scale leaves.

CONES terminal and curved down on short stalks, berrylike, round, 1/4–5/16 in. (6–8 mm) in diameter, light blue and covered with a bloom, fleshy, resinous. **Seeds** 4 or fewer, brown, 1/8 in. (3 mm) long.

HABITAT Rare and local on dry rocky slopes and sunny sands.

USES Used as an ornamental ground cover in interior and south central Alaska.

CREEPING JUNIPER

COMMON JUNIPER

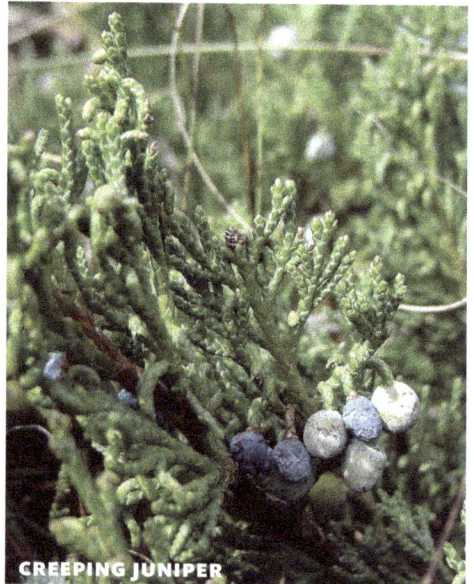

CREEPING JUNIPER

ARBORVITAE *Thuja*

Western Redcedar
Thuja plicata Donn ex D. Don

OTHER NAMES giant arborvitae, canoe cedar, shinglewood, Pacific redcedar, arborvitae

DESCRIPTION Large evergreen tree 70–100 feet (21–30 m) tall, sometimes 130 feet (40 m), with tapering trunk 2–4 feet (0.6–1.2 m) in diameter, sometimes to 6 feet (1.8 m), swollen or buttressed base, pointed conical crown, and horizontal branches curving upward at tips.

LEAVES scalelike, flattened, 1/16–1/8 in. (1.5–3 mm) long, on leader twigs to 1/4 in. (6 mm) long and pointed, shiny yellow green above and dull green below.

LEAFY TWIGS flattened, in fanlike sprays, slightly drooping, older twigs gray and smooth.

BARK gray or brown, thin, fibrous and stringy or shreddy, becoming thick and furrowed into long ridges.

WOOD with the distinctive odor of cedars, fine-textured, straight-grained, lightweight, moderately soft, brittle. Heartwood reddish brown, the narrow sapwood white.

CONES clustered near ends of twigs and becoming turned up on short stalks, elliptic, 1/2 in. (12 mm) long, light brown, composed of several paired elliptic leathery cone-scales. **Seeds** 3 or fewer under a cone-scale, 3/16 in. (5 mm) long, light brown, with 2 narrow wings.

HABITAT Western redcedar is native in the southern half of southeast Alaska from sea level to 3,000 feet altitude on the west slopes of the Coast Range, attaining its largest size below 500 feet. Although sometimes in pure stands, it is also dominant in the redcedar-hemlock forest and scattered in the hemlock-spruce forest. It is of moderately slow growth and long-lived.

USES Western redcedar is well suited for boat and canoe construction. It is the most widely used wood for shingles. Other uses of this very durable, lightweight wood are utility poles, fenceposts, light construction, pulp, clothes closets and chests, conduits, piling, and fish-trap floats. Alaska Natives employed the wood for totem poles, dugout canoes, and houses, and made mats, baskets, and ropes from the stringy bark. This is an important timber tree of the coast region of British Columbia. Western redcedar is exported to Japan in log form, though some is used locally.

WESTERN REDCEDAR

WESTERN REDCEDAR

WESTERN REDCEDAR - FEMALE CONES

■ PINE FAMILY *Pinaceae*

These cone-bearing trees are resinous softwoods with needlelike or scalelike evergreen leaves, with seeds exposed in cones, these usually hard and woody. Pollen borne in small male cones, usually on the same plant; true flowers and fruits are lacking. Alaska's sole conifer in the yew family (*Taxus*) has seeds borne singly in a scarlet, juicy, cuplike disk, rather than in a cone.

The pine family (Pinaceae) is well-represented in Alaska by 5 genera and 9 species of trees with narrow, mostly long needles. The cones have many cone-scales, each bearing two long-winged seeds at its base. Characters of the Alaska species are:

• **Larch** (*Larix*), the only Alaska conifer shedding its leaves in fall and leafless in winter. One species, tamarack (*L. laricina*), with slender flexible needles borne 12–20 in a cluster on short stout spur twigs (or single on leading twigs).

• **Pine** (*Pinus*), 1 species, lodgepole pine (*P. contorta*), with 2 varieties. Needles 2 in a bundle or cluster with sheath at base, relatively long and stiff. Cones one-sided, with many prickly cone scales.

• **Spruce** (*Picea*), 3 species, black, white, and Sitka spruce. Needles sharp-pointed and stiff, either 4-angled or flattened and slightly keeled, extending out on all sides of twig. There is no leafstalk, but each leaf is attached on a small stalklike or peglike projection of the twig. Older twigs without needles are rough because of these projections. Cut branches of spruce and hemlock shed their needles promptly upon drying. The cones hang down. (If preparing botanical specimens, immersing of freshly cut twigs in boiling water for a few minutes before pressing reduces shedding of needles.)

TAMARACK

• **Hemlock** (*Tsuga*), 2 species, western and mountain hemlock. Needles short, blunt, soft and not stiff, flat or slightly keeled, with short leafstalks, spreading in 2 rows or curved upward. As in spruce, the older twigs are slightly rough from the peglike projections. The cones hang down.

• **Fir** (*Abies*), 2 species, Pacific silver fir and subalpine fir. Needles flat and without leafstalks, often spreading in 2 rows or curving upward. Older twigs smooth with round leaf-scars. Cones upright in highest branches of the narrow pointed crowns. As the cone-scales fall from the axis at maturity, old cones are not found on or under the trees.

Conifers, or softwoods, are economically the most important group of trees in Alaska. Many have tall straight trunks and narrow crowns, except where dwarfed near the limits of tree growth (however, the two native junipers are low shrubs). These narrow-leaf evergreens make up nearly all the trees of the coastal forests of southeast Alaska and most of the timber of the interior forests. They furnish almost all of the state's lumber, pulp-wood, building logs, and other wood products.

FIR *Abies*

Trees with narrow, pointed crowns and mostly horizontal branches. **Leaves** are flat needles without leafstalks, those on lower branches often spreading in 2 rows along the twig, others mostly curving upward. Older twigs are smooth with round leaf-scars. **Cones** are upright and stalkless in the highest branches. At maturity the cone-scales and seeds are shed, but the narrow upright axis persists on the twig; no old cones remain on the trees or on the ground. Two species of fir are present in southeastern Alaska, but are not common.

KEY TO ALASKA FIRS

1 Leaves (needles) shiny dark green on upper surface and silvery white with many lines (stomata) on lower surface **PACIFIC SILVER FIR** (*Abies amabilis*)

1 Leaves (needles) dull dark green with whitish lines (stomata) on both surfaces **SUBALPINE FIR** (*Abies lasiocarpa*)

Pacific Silver Fir

Abies amabilis (Dougl. ex Loud.) Dougl. ex Forbes

OTHER NAMES silver fir, white fir (lumber)

DESCRIPTION Medium-sized resinous and aromatic tree rare and local in extreme southeast Alaska, becoming 80 feet (24 m) tall and 24 in. (60 cm) in trunk diameter, maximum 149 feet (45 m) tall and 49 in. (1.24 m) in diameter.

LEAVES (needles) crowded and spreading, stalkless, ¾ –1¼ in. (2–3 cm) long, flat, deeply grooved and shiny dark green above, beneath silvery white with whitish lines (stomata), those on lower branches notched or rounded at tips and spreading in 2 rows, those toward top of tree shorter and sharp-pointed, twisted in brushlike mass on upper side of twig.

TWIGS slender, finely hairy.

BARK smooth, gray, splotched with white.

WOOD with whitish sapwood and pale brown heartwood, fine-textured, lightweight, soft.

CONES in highest branches, upright, 4–5 in. (10–12.5 cm) long, 2–2½ in. (5–6 cm) in diameter, purplish, finely hairy or nearly hairless; many fan-shaped rounded overlapping scales, falling from axis in autumn. **Seeds** light brown, about 1 in. (2.5 cm) long, including broad wing.

HABITAT Rare and local in extreme southeast Alaska. It has been recorded from well-drained lower slopes of canyons, benches, and flats from sea level to 1,000 feet altitude. In the Salmon River valley near Hyder, it is common in the coastal forest of Sitka spruce and eastern hemlock, being very shade tolerant. South through Pacific coast region of British Columbia and in mountains to Oregon and northwestern California.

PACIFIC SILVER FIR

USES The trees are logged with other conifers. Fir logs are sawed into lumber with Sitka spruce, if large and clear, or chipped with hemlock and used for pulp. Southward, where more abundant, the wood is used for interior finish.

PACIFIC SILVER FIR

PACIFIC SILVER FIR - CONES

Subalpine Fir

Abies lasiocarpa (Hook.) Nutt.

OTHER NAMES alpine fir, white fir (lumber)

DESCRIPTION Small to medium-sized evergreen tree, rare and local in southeast Alaska, commonly 20–60 feet (6–18 m) high and 4–12 in. (10–30 cm) in trunk diameter, with long, narrow, sharp-pointed or spirelike crown and branches extending nearly to base, resinous and aromatic. However, larger trees to 95 feet (29 m) tall and 27 in. (69 cm) in diameter have been observed.

LEAVES (needles) crowded and spreading, stalkless, 3/4–1½ in. (2–4 cm) long, flat, dark blue green and with whitish lines (stomata) on both sides, grooved above, those on lower branches rounded or occasionally notched at tip and in 2 rows, those near top of tree shorter, pointed, stiff, and twisted upward and curved on upper side of twig.

TWIGS gray, rusty hairy.

BARK ash gray, smooth, thin.

WOOD pale brown, fine-textured, lightweight, soft, usually knotty because of the many persistent branches.

CONES in highest branches, upright, cylindric, 2½–4 in. (6–10 cm) long and 1¼–1½ in. (3–4 cm) in diameter, dark purple, finely hairy; many fan-shaped, rounded, overlapping scales, falling from axis in autumn. **Seeds** light brown, 5/8 in. (1.5 cm) long, including broad wing.

HABITAT Cool, moist subalpine slopes near timberline, where becoming shrubby or prostrate, also found on the valley floors; it appears to be very shade tolerant. In southeast Alaska, subalpine fir from the interior of British Columbia crosses over the divide of the Coast Range, and extends from the Canadian border down to sea level, scattered with Sitka spruce, hemlock, and black cottonwood. Near Skagway, subalpine fir descends from timberline at 3,000 feet (914 m) to sea level. Northeastward in Yukon Territory, this species occurs within 125 miles (200 km) of the Alaska border along Stewart River, a tributary of the Yukon River. Subalpine fir has also been reported from several localities in south central Alaska.

SUBALPINE FIR

SUBALPINE FIR

SUBALPINE FIR

SUBALPINE FIR

SUBALPINE FIR - BARK

LARCH *Larix*

Tamarack
Larix laricina (Du Roi) K. Koch

OTHER NAMES Alaska larch, eastern larch, hackmatack

SYNONYM *Larix alaskensis* W.F. Wight

DESCRIPTION Small to medium-sized deciduous tree 30–60 feet (9–18 m) high, with straight tapering trunk 4–10 in. (10–25 cm) in diameter, occasionally to 75 feet (24 m) tall and 13 in. (33 cm) in diameter, horizontal branches extending nearly to ground, and thin pointed crown of blue green foliage.

LEAVES (needles) shedding in fall (deciduous), in crowded clusters of 12–20 on short stout spur twigs or branches or single on leading twigs, 3/8–1 in. (1–2.5 cm) long, very narrow, slender and flexible, 3–angled, blue green, turning yellow before falling in early autumn.

TWIGS long, stout, dull tan, hairless; with many short stout spur twigs to 1/4 in. (6 mm) long, bearing crowded raised leaf-scars, becoming blackish and rough.

WINTER BUDS small, round, about 1/16 in. (2 mm) long, covered by many short-pointed overlapping scales.

BARK dark gray, smoothish, thin, becoming scaly and exposing brown beneath.

WOOD light brown, hard, heavy, resinous.

CONES curved upright on short stalks along horizontal twigs, rounded, 3/8–5/8 in. (1–1.5 cm) long, dark brown, composed of about 20 rounded, finely toothed cone-scales, opening in early autumn and remaining attached in winter. **Seeds** light brown, 1/2 in. (12 mm) long, including long broad wing.

HABITAT Muskegs and various moist soils of the interior in open stands with paper birch, black spruce, alders, and willows. Occasionally it forms dense stands on floodplains with black spruce and white spruce. Where it does occur naturally on upland well drained sites, its growth rate may be equal to that of white spruce; one stand in the Tanana Valley has produced trees 13 in. (33 cm) in diameter in 100 years. In Alaska, tamarack is restricted to drainages between Brooks Range on the north and Alaska Range on the south.

USES The durable, strong wood is used to some extent for poles, railroad ties, and fenceposts.

TIP The soft needles arranged in bundles distinguish tamarack from both white spruce (*Picea glauca*) and black spruce (*P. mariana*). In winter, needle-less trees of tamarack can be distinguished from dead spruce trees by the presence of short spur-like twigs.

TAMARACK

TAMARACK - CONES

TAMARACK

SPRUCE *Picea*

Spruce trees have short leaves (needles) spreading on all sides of twig, the **needles** mostly 4-sided or slightly flattened, sharp-pointed and stiff, shedding promptly on drying. **Twigs** are rough from the peglike bases of leaves. The **cones** hang down.

KEY TO ALASKA SPRUCE

1 Leaves (needles) flattened but slightly keeled, with 2 whitish bands (stomata) on lower surface; twigs hairless; cones cylindric, 2–3½ in. (5–9 cm) long, falling at maturity

SITKA SPRUCE (*Picea sitchensis*)

1 Leaves (needles) 4-angled, with whitish lines (stomata) on all sides   **2**

2 Twigs hairy; needles mostly less than 1/2 in. (12 mm) long, resinous; cones egg-shaped or nearly round, mostly less than 1 in. (2.5 cm) long, curved down on short stalks, remaining on tree

BLACK SPRUCE (*Picea mariana*)

2 Twigs hairless; needles more than 1/2 in. (12 mm) long, with skunklike odor when crushed; cones cylindric, 1¼ –2½ in. (3–6 cm) long, falling at maturity

WHITE SPRUCE (*Picea glauca*)

White Spruce

Picea glauca (Moench) Voss

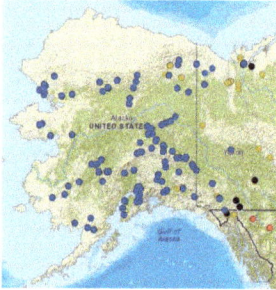

OTHER NAMES western white spruce, Canadian spruce, Alberta spruce

DESCRIPTION White spruce, the most important tree of the spruce-birch interior forest, is a medium-sized to large tree 40–70 feet (12–21 m) high and 6–18 in. (15–46 cm) in trunk diameter. On the best sites it reaches 80–115 feet (24–35 m) and 30 in. (76 cm), but at timberline it becomes a prostrate shrub with a broad base below the snow-cover line. Crown pointed and usually very narrow and spirelike, sometimes broad and conical, composed of slightly drooping branches with upturned ends and many small drooping side twigs.

LEAVES (needles) short-stalked, spreading on all sides of twig but massing on top near ends, 1/2–3/4 in. (12–20 mm) long, 4-angled, sharp-pointed, stiff, blue green, with whitish lines on all sides; leaves and twigs with skunk-like odor when crushed.

TWIGS slender, hairless, orange brown, becoming rough from peglike bases of leaves.

BARK thin, gray, smoothish or in scaly plates, the cut surface of inner bark whitish.

WOOD almost white, the sapwood not easily distinguished, moderately lightweight, moderately soft, of fine and moderately uneven texture, with growth rings easily seen in cross-sections.

WHITE SPRUCE

CONES nearly stalkless, hanging down, cylindric, $1\frac{1}{4}$ –$2\frac{1}{2}$ in. (3–6 cm) long, shiny light brown, falling at maturity; cone-scales thin and flexible, margins nearly straight and without teeth. **Seeds** brown, about 3/8 in. (10 mm) long, including large wing.

HABITAT White spruce is the commonest tree of interior Alaska, occuring from near sea level to tree-line at about 1,000–3,500 feet (305–1,607 m). The tree-line is lowest in the north and west and on north-facing slopes and highest in the southeast interior and on south-facing slopes. This species is found in mostly open forests, usually with paper birch or in pure stands. In a few places, it extends to tidewater. Although not exacting as to site, this species grows best on well drained soils on south-facing gentle slopes and sandy soils along the edges of lakes and rivers. It forms the tallest forests along the large rivers, where running water thaws the soil. It is seldom found where permafrost is close to the surface. White spruce often replaces balsam poplar along the river floodplains and also invades the open forests of birch and aspen that follow fire. The trees have average growth rate, attaining an age of 100–200 years at maturity.

USES White spruce is used extensively in interior Alaska for cabin logs, peeled and in natural form, sawed flat on 3 sides, or milled on lathes into uniformly round logs having diameters of 6, 8, or 10 inches (15, 20, or 25 cm). Large numbers of pilings and rough timbers from interior Alaska have been transported to the North Slope for construction of oil drilling platforms. Timbers for bridges and corduroy roads are other uses. A small quantity is cut for fuel also. This species supplies much of the lumber sawed in interior Alaska, also dimension material for buildings in light and medium construction. Early uses included flumes, sluice boxes, and boats.

In Canada, white spruce is the most important commercial tree species and the foremost pulpwood. Uses include scaffolding planks, paddles and oars, sounding boards in musical instruments, shop fittings, agricultural implements, kitchen cabinets, boxes, cooperage, shelving, veneer, and plywood. The seasoned wood is almost tasteless and odorless and well suited for food containers.

NOTE Alaska trees commonly have very narrow crowns and short broad cones. Another western variation scattered in Alaska has smooth bark with resin blisters (as in fir) and relatively broad crown. On Kenai Peninsula, where this species meets Sitka spruce, hybrids or intermediate trees occur, as noted under that species.

WHITE SPRUCE

WHITE SPRUCE

WHITE SPRUCE

WHITE SPRUCE – BARK

Black Spruce
Picea mariana (P. Mill.) B.S.P.

OTHER NAMES bog spruce, swamp spruce

DESCRIPTION Evergreen resinous tree of interior forests, usually small and 15–30 feet (4.5–9 m) high, and 3–6 in. (7.5–15 cm) in trunk diameter, with narrow pointed crown. Often a shrub 10 feet (3 m) or less in height. Sometimes a medium-sized tree to 50–60 feet (15–18 m) tall and 9 in. (23 cm) in trunk diameter, the maximum height measured 72 feet (22 m). The branches are short, sparse, and often slightly drooping at ends.

LEAVES (needles) short-stalked, spreading on all sides of twig, 1/4–5/8 in. (6–15 mm) long, 4-angled, pointed, stiff, ashy blue green, with whitish lines (stomata) on all sides.

TWIGS slender, hairy, covered with very short reddish hairs, becoming brown and rough from peglike bases of leaves.

BARK thin, composed of gray or blackish scales, brown beneath, the cut surface of inner bark yellowish.

WOOD yellowish white, light-weight, soft, fine-textured, with growth rings very narrow to almost microscopic.

CONES curved downward on short stalks, small and short, egg-shaped or nearly round, 5/8–1¼ in. (1.5–3 cm) long, dull gray or blackish, remaining on tree several years and often conspicuously clustered in tree tops; cone-scales rigid and brittle, rounded, and slightly toothed. **Seeds** brown, about 1/2 in. (12 mm) long including large wing.

HABITAT Black spruce is characteristic of cold wet flats, muskegs, north-facing slopes, silty valley terraces, and lake margins in the spruce-birch interior forests up to an altitude of 2,000 feet, locally to 2,700 feet. Extending to treeline on gentle damp slopes, such as the northern side of the Alaska Range. Dense pure stands often develop following fires in wet areas. Interior Alaska north to southern slopes of Brooks Range but at lower elevations and not as far north as white spruce.

USES The wood is of slight importance for lumber because of the small size of the trees. Occasionally the logs are cut along with white spruce for cabins. The trees are important as fuel, especially in stands killed by fire, remaining standing and well preserved for several decades. Southward black spruce is a popular Christmas tree.

NOTE Besides its usually different habitat and smaller size with more compact branching, black spruce is distinguished from white spruce by the shorter

BLACK SPRUCE

and blunter needles, hairy twigs, and smaller cones with brittle, slightly toothed cone-scales curved down on short stalks and remaining attached several years. The twigs of black spruce are reported to be tougher and gummier also. These two species of the interior forests can be distinguished also in the seedling stage by the finely toothed leaf margins in white spruce and absence of teeth in black spruce. Logs and tree trunks can be identified by inner bark color, yellowish in black spruce and whitish in white spruce. Annual growth rings of black spruce wood are also narrower .

BLACK SPRUCE - NEEDLES

BLACK SPRUCE

BLACK SPRUCE - CONES

Sitka Spruce

Picea sitchensis (Bong.) Carr.

OTHER NAMES tideland spruce, yellow spruce, western spruce, silver spruce, coast spruce

DESCRIPTION Large to very large evergreen tree to 160 feet (49 m) in height and 3–5 feet (0.9–1.5 m) in trunk diameter, infrequently to 200–225 feet (61–69 m) and 7–8 feet (2.1–2.4 m) or more. From the much enlarged or buttressed base, the tall straight evenly tapering trunk rises to an open, pointed, broad, conical crown with horizontal branches.

LEAVES (needles) standing out on all sides of twig, flattened and slightly keeled, 5/8–1 in. (15–25 mm) long, sharp-pointed, dark green, the upper surface slightly keeled or angled and with 2 whitish bands (stomata), lower surface rounded or slightly keeled and sometimes with few whitish lines.

TWIGS stout, stiff, hairless, light brown to dark brown, becoming rough from peglike bases of leaves.

BARK gray and smoothish on small trunks, thin, becoming dark purplish brown with scaly plates, the inner bark whitish with brown dots.

WOOD moderately lightweight, moderately soft, of fine and moderately even texture, and usually very straight grained. Sapwood nearly white and heartwood light reddish brown.

CONES short-stalked, hanging down, cylindric, 2–3½ in. (5–9 cm) long, light orange brown, falling at maturity; cone-scales long, stiff, thin, rounded and irregularly toothed. **Seeds** brown, about 1/2 in. (12 mm) long, including large wing.

HABITAT Sitka spruce forms more than 20 percent of the hemlock-spruce coastal forests of Alaska and also occurs in pure stands. It grows more rapidly and to larger size than western hemlock and is more light-requiring. The largest old growth trees in southeast Alaska have trunk diameters exceeding 8 feet (2.4 m) and ages of 500–750 years or more. Many years ago there was reported a giant 14½ feet (4.4 m) in trunk diameter measured 6 feet (1.8 m) above the ground, but further information including the locality is lacking.

At Afognak and Kodiak Islands there are pure stands of Sitka spruce, the only conifer. On Kodiak Island near the southwestern limit, this tree is reported to be migrating westward during the past few centuries.

This species extends from sea level to the timberline up to about 3,000 feet in the coastal mountains but grows mainly at altitudes below 1,500 feet. However, dwarf plants have been seen as high as 3,500–3,900 feet on unglaciated rocky outcrops (nunataks) projecting above the Juneau Ice Field.

In bare or open areas, such as at Glacier Bay, the bushy trees often propagate by layering. The lowest branches touch the ground, become partly covered, develop roots, and then turn upward to form separate trees. Sprouts from stumps have been observed also.

Small groves of Sitka spruce trees were planted as early as 1805 by Russians at Unalaska, near the eastern end of the tree-

SITKA SPRUCE

less Aleutian Islands and far outside the tree limits. These trees are still growing and have produced cones. Younger trees are absent, perhaps because of grazing. Several plantings have been made also on other Aleutians. Both the common and scientific names honor Sitka Island, now Baranof Island, where the southeast Alaska town of Sitka is located.

USES Sitka spruce is the largest and one of the most valuable trees in Alaska, and also the state tree. It produces high-grade wood pulp, considered the best on the Pacific coast. The wood, with that of western hemlock, is used extensively in manufacturing newsprint. This species is also the principal sawtimber tree of southeast Alaska, made into all the usual forms of lumber.

The high-grade lumber from the large clear trunks has many uses. It is the most important wood for airplane and glider construction, and in World War II was used especially in British mosquito bombers. Other important uses are oars, ladders and scaffolding, and boats, particularly racing sculls. Resonant qualities, large size, and uniformity make the wood valuable for piano sounding boards. The low-grade lumber is made into packing boxes for the Alaska salmon industry. Other uses are general construction, food containers, shelving, and kitchen furnishings. Alaska has about 40 percent of the total supply of this species and over 85 percent of the United States supply.

NOTE On Kenai Peninsula there are natural hybrids between white spruce and Sitka spruce (*Picea glauca* x *sitchensis*; Lutz spruce, *Picea* x *lutzii* Little). The hybrid is a tree 55–70 feet (17–21 m) high and 1–1½ feet (30–45 cm) in trunk diameter. Hybrid trees are recognized by their leaves and cones intermediate between those of the parent species. The leaves are slightly 4-angled, less so than in white spruce and are near Sitka spruce in the whitish upper surfaces. The cones are intermediate in size or small as in white spruce. Cone-scales are short as in white spruce but like Sitka spruce in being thin, light brown, and irregularly toothed. These hybrids are found on Kenai Peninsula where the ranges of the two species meet and overlap slightly and may be sought elsewhere along the border between the coastal and interior forest types.

SITKA SPRUCE

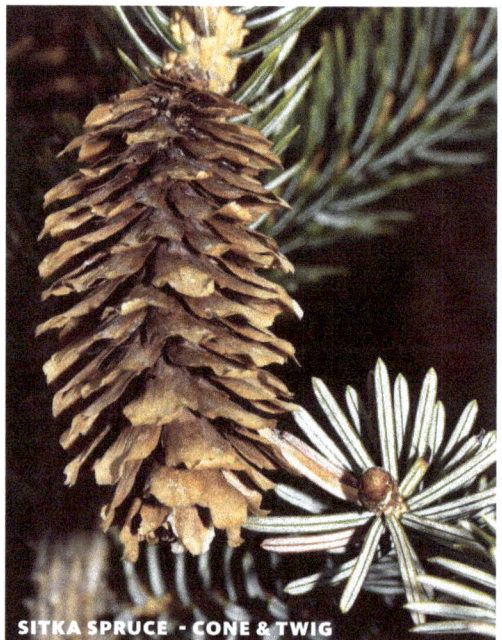

SITKA SPRUCE · CONE & TWIG

SITKA SPRUCE

SITKA SPRUCE - YOUNG CONES

PINE *Pinus*

Lodgepole Pine
Pinus contorta Dougl. ex Loud.

OTHER NAMES scrub pine, tamarack pine

NOTE The general description and range of this species are followed by similar notes for the 2 varieties in Alaska.

DESCRIPTION Small to large evergreen, resinous tree of southeast Alaska, 20–75 feet (6–23 m) tall and 8–32 in. (20–81 cm) in trunk diameter, with crown rounded spreading or narrow pointed.

LEAVES (needles) 2 in a bundle with sheath at base, 1–2 1/4 in. (2.5–6 cm) long, relatively long and stiff, often twisted, yellow green to dark green with whitish lines (stomata).

TWIGS stout, orange when young, becoming gray brown and rough. **Winter buds** short-pointed, of many narrow red brown scales.

BARK gray to dark brown, scaly, thin or becoming thick.

WOOD resinous or pitchy, coarse-textured, straight-grained (scrubby trees with spiral grain), moderately lightweight, moderately soft. Heartwood light yellow to yellow brown, sapwood narrow and whitish.

CONES 1 to few, almost stalkless, egg-shaped, one-sided, $1\frac{1}{4}$–2 in. (3–5 cm) long, light yellow brown, with many prickly cone scales, maturing in 2 years, persistent, opening or remaining closed many years. **Seeds** brown, about 5/8 in. (15 mm) long, including the long broad wing.

USES Wood of lodgepole pine of the Rocky Mountain region is suitable for pulping for papers and fiber-board. Other uses are lumber, railroad ties, mine timbers, and poles, posts, and fuelwood. The lumber is mostly for rough construction, occasionally for boxes, siding, finish, and flooring.

NOTE This species, including 3 geographic varieties, has a broad range from southeast Alaska, central Yukon, and southwestern Mackenzie, south in mountains and along coast to Colorado, Utah, and California; also local in northern Baja California. Alaska's only native species of pine is not important for lumber because of its mostly small size and limited occurrence. The wood is used for poles and fuel. The sweet orange-flavored sap served Alaska Natives as a delicacy, fresh or dried. In the vicinity of Fairbanks, the inland variety has been introduced as a fast growing, hardy shade tree.

LODGEPOLE PINE

1 Cones pointing backward, opening at maturity; generally low spreading tree of muskegs in coastal forests   SHORE PINE (*Pinus contorta* var. *contorta*)

1 Cones pointing outward, mostly remaining closed many years; tree often tall and narrow, of inner fiord forests at head of Lynn Canal (Skagway to Haines)
LODGEPOLE PINE (*Pinus contorta* var. *latifolia*)

Shore Pine
Pinus contorta Dougl. var. *contorta*

OTHER NAMES lodgepole pine, scrub pine, tamarack pine

DESCRIPTION Shore pine, the common pine through southeast Alaska, is often a low spreading or scrubby tree 20–40 feet (6–12 m) high and 8–12 in. (20–30 cm) in trunk diameter; however, it sometimes grows to 75 feet (23 m) tall and 18–32 in. (45–81 cm) in diameter.

CONES point backward on the twigs, opening at maturity in October-November but remaining attached.

HABITAT The dwarf coastal form occurs in open muskegs of peat moss and on benches near lakes. Intolerant of shade, it grows in open stands as a scrub pine, straight when young but gnarled in age, with large branches extending almost to the ground. On the poorest sites, it is often like a prostrate shrub. Trees are best developed and largest in the better-drained borders between muskeg and hemlock or hemlock-redcedar stands. Occasionally the trees are rapidly growing pioneers after infrequent fires or logging, or on outwash sand and gravel.

This coastal variety ranges throughout southeast Alaska north to the head of Lynn Canal at Haines and to Glacier Bay and Dixon Harbor. The northwestern outlier is an area of several square miles on rolling muskegs about 15 miles (24 km) east of Yakutat, where the trees of poor form reach 40 feet (12 m) in height and 1 feet (30 cm) in trunk diameter.

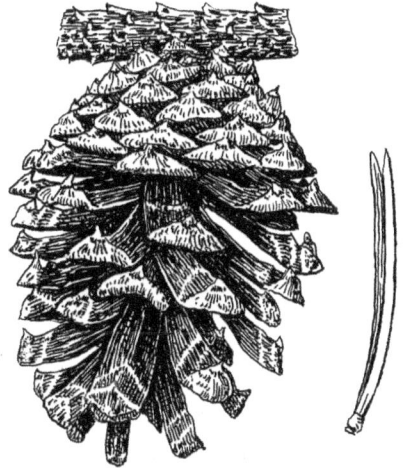

LODGEPOLE PINE

Lodgepole Pine
Pinus contorta var. *latifolia* Engelm.

OTHER NAME Rocky Mountain lodgepole pine

DESCRIPTION in Alaska, this mostly tall form with narrow crown becomes 50–75 feet (15–23 m) high and 8–12 in. (20–30 cm) in trunk diameter, and somewhat larger southward.

CONES hard, heavy, pointing outward, mostly remaining closed many years, opening after a forest fire to release seeds (however, in Alaska some cones open at maturity).

HABITAT This variety of lodgepole pine reportedly crosses the Coast Range from Canada into Alaska only in the vicinity of Skagway and Haines near the northernmost end of southeast Alaska. It forms stands in the mixed forest with Sitka spruce, western paper birch, and subalpine fir (also from the Rocky Mountains) and in the inner fiords down to sea level.

NOTE This inland variety differs from shore pine in being generally a taller tree with narrow crown and thinner scaly bark, in having slightly longer needles, and in the slightly larger, heavier, closed cones which point outward on the twig rather than backward.

LODGEPOLE PINE - CLOSED CONE

LODGEPOLE PINE - OPEN CONE

HEMLOCK *Tsuga*

Hemlock trees have very slender leading twigs or leaders which are curved down or nodding. The **leaves** are short needles, flat or half-round, blunt, soft, and not stiff, with short leafstalks, shedding promptly on drying. **Twigs** are very slender, becoming roughened by peglike bases.

KEY TO ALASKA HEMLOCK

1   Leaves (needles) flat, appearing in 2 rows, shiny dark green above, with 2 whitish bands (stomata) on lower surface; 5/8–1 in. (1.5–2.5 cm) long
WESTERN HEMLOCK (*Tsuga heterophylla*)

1   Leaves (needles) half-round and keeled or angled beneath, crowded on all sides of short side twigs, blue green, with whitish lines (stomata) on both surfaces; cones cylindric, 1–2½ in. (2.5–6 cm) long   MOUNTAIN HEMLOCK (*Tsuga mertensiana*)

Western Hemlock

Tsuga heterophylla (Raf.) Sarg.

OTHER NAMES Pacific hemlock, west coast hemlock (lumber)

DESCRIPTION Large evergreen tree becoming 100–150 feet (30–46 m) tall and 2–4 feet (0.6–1.2 m) in trunk diameter, with long slender trunk often becoming fluted when large, and short narrow crown of horizontal or slightly drooping branches, the very slender leading twig curved down or nodding. The largest trees are as much as 190 feet (58 m) in height and 5 feet (1.5 m) or more in diameter.

LEAVES (needles) short-stalked, spreading in 2 rows, 1/4–7/8 in. (6–22 mm) long, flat, rounded at tip, flexible, shiny dark green above, and with 2 whitish bands (stomata) on lower surface.

TWIGS slender, dark reddish brown, finely hairy, roughened by peglike bases after leaves fall.

BARK reddish brown to gray brown, becoming thick and furrowed into scaly plates; a pocket-knife will disclose the red inner bark not found in spruce.

WOOD moderately light-weight, moderately hard, of moderately fine and even textured, nonresionous. Heartwood pale reddish brown, sapwood similar or whitish.

CONES stalkless and hanging down at end of twig, small, elliptic, 5/8–1 in. (1.5–2.5 cm) long, brown, with many thin papery scales. **Seeds** about 1/2 in. (12 mm) long (including large wing).

HABITAT Western hemlock occurs in the hemlock-spruce coastal forests of southeast and southern Alaska, but does not go as far west as Sitka spruce.

WESTERN HEMLOCK

NOTE Western hemlock is the most abundant and one of the most important tree species in southeast Alaska and forms more than 70 percent of the dense hemlock-spruce coastal forests. This species attains its largest size on moist flats and lower slopes, but with abundant moisture, both atmospheric and soil, it grows well on shallow soils. It is very tolerant of shade.

This species is one of the best pulpwoods for paper and paper-board and products such as rayon. Other important uses are lumber for general construction, railway ties, mine timbers, and marine piling. The wood is suited also for interior finish, boxes and crates, kitchen cabinets, flooring and ceiling, gutter stock, and veneer for plywood. The outer bark contains a high percentage of tannin and is a potential source of this product. Alaska Natives made coarse bread from the inner bark of this tree and shore pine. Western hemlock is the state tree of Washington.

WESTERN HEMLOCK - CONE

WESTERN HEMLOCK

Mountain Hemlock

Tsuga mertensiana (Bong.) Carr.

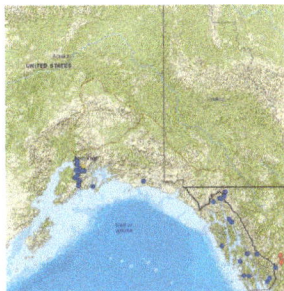

OTHER NAMES alpine hemlock, black hemlock

DESCRIPTION Small to large evergreen tree becoming 50–100 feet (15–30 m) high and 10–30 in. (25–76 cm) in trunk diameter, maximum about 125 feet (38 m) and 40 in. (1 m), with marked taper when open grown, narrow crown of horizontal or drooping branches, and very slender leading twig curved down or nodding; a shrub near timber line.

LEAVES (needles) mostly crowded on all sides of short side twigs and curved upward, short-stalked, 1/4–1 in. (6–25 mm) long, flattened above and rounded, keeled, or angled beneath (half-round in section), stout and blunt, blue green and with whitish lines (stomata) on both surfaces.

TWIGS mostly short, slender, light reddish brown, finely hairy, roughened by peglike bases after leaves fall.

BARK gray to dark brown, thick, and deeply furrowed into scaly plates.

WOOD moderately heavy, moderately hard, and moderately fine and even textured. Heartwood pale reddish brown, sapwood thin and similar or whitish.

CONES stalkless and usually hanging down, cylindric, 1–2½ in. (2.5–6 cm) long and 3/4 in. (2 cm) wide, purplish but turning brown, with many thin papery scales; very resinous when young. **Seeds** light brown, about 1/2 in. (12 mm) long including large wing.

HABITAT Southeast and southern Alaska, extending from sea level to an altitude of 3,000–3,500 feet, growing at an altitude higher than other trees. On upland sites, it is well formed and resembles western hemlock. Toward timberline, it replaces the latter and becomes a prostrate shrub. It grows with shore pine in muskegs of deep peat, as well as on subalpine slopes on the ocean side of the Coast Range in southeast Alaska. In the Prince William Sound and Cook Inlet regions, mountain hemlock is found on better drained slopes and near tidewater, and reaches its maximum height.

USES The wood is marketed with western hemlock, being similar but somewhat more dense, and has the same uses. Nearly pure stands of mountain hemlock on Prince of Wales Island have been logged for pulp. The wood has been used for railroad ties. However, in the higher altitudes where commonly found, mountain hemlock is largely inaccessible and unimportant commercially.

MOUNTAIN HEMLOCK

ETYMOLOGY This species honors the German naturalist Karl Heinrich Mertens (1796–1830), who discovered it near Sitka, Alaska, in 1827.

ADDITIONAL SPECIES Douglas-fir (*Pseudotsuga menziesii* (Mirb.) Franco) though not native to Alaska, is sometimes planted in the southeast as an ornamental and in forestry tests; growth is rapid. The flat leaves (needles) are 5/8–1¼ in. (1.5–3 cm) long, and resemble those of fir but are narrowed into stalks at their base and have an elliptic leaf-scar. Winter buds are pointed, red brown, and not resinous. The elliptic, light brown cones, 2–3½ in. (5–9 cm) long, hang down and have thin rounded cone-scales and prominent 3-toothed bracts.

Douglas-fir, one of the world's most valuable timber trees, is widespread in the Pacific coast and Rocky Mountain regions north in British Columbia nearly to Alaska. On the coast it extends to about the north end of Vancouver Island. In interior British Columbia, it extends north to Tacla Lake at latitude 55°, north of the southern tip of Alaska.

DOUGLAS-FIR · CONE

MOUNTAIN HEMLOCK - YOUNG CONES

■ YEW FAMILY *Taxaceae*

Pacific Yew
Taxus brevifolia Nutt.

OTHER NAME western yew

DESCRIPTION Small tree or large shrub of extreme south end of southeast Alaska, to 20–30 feet (6–9 m) tall, with straight conical trunk 2–6 in. (5–15 cm) or rarely 12 in. (30 cm) in diameter at breast height, with open crown or horizontal or drooping branches.

LEAVES (needles) in 2 rows, 1/4–3/4 in. (12–20 mm) long, flat, slightly curved, stiff or soft, abruptly pointed but not prickly, shiny yellow green above, pale green beneath, not resinous. Petioles yellow, extending down the slender twigs, twisting to produce an even, comblike arrangement of needles.

BARK purplish brown, thin, scaly, ridged, and fluted.

WOOD bright red with thin light yellow sapwood, fine-textured, heavy, hard, elastic.

FLOWERS Pollen and seeds on different trees (dioecious). Seeds single, 3/8 in. (1 cm) long, brown, exposed at tip but partly surrounded by a thick scarlet, juicy, cuplike disk or "berry."

HABITAT Uncommon in the south end of southeast Alaska, near sea level on poor sites and in canyons. It is scattered in the understory of the coast forest of western redcedar, western and mountain hemlock, and Sitka spruce. The irregular distribution may be related to dispersal of the seeds by birds. Growth is slow.

USES Southward, the strong, durable wood is used for poles, bows, canoe paddles, and cabinet work. However, in Alaska the trees are too scarce to be commercially important.

NOTE The seeds are poisonous when eaten, causing vomiting, diarrhea, and inflammation of urinary ducts and the uterus. Also, yew foliage is poisonous when browsed by livestock. However, the juicy scarlet "berries" around the seeds are not toxic.

PACIFIC YEW

PACIFIC YEW

ALASKA TREES & SHRUBS—ANGIOSPERMS

■ MUSKROOT FAMILY *Adoxaceae*

Our 2 shrubby genera, *Sambucus* and *Viburnum,* were previously included in the honeysuckle family (Caprifoliaceae).

KEY TO ALASKA ADOXACEAE SHRUBS

1 Leaves pinnately compound	**PACIFIC RED ELDER** (*Sambucus racemosa*)
1 Leaves simple, lobed	**SQUASHBERRY** (*Viburnum edule*)

Pacific Red Elder
Sambucus racemosa L.

OTHER NAMES scarlet elder, redberry elder, stinking elder, elderberry

SYNONYM *Sambucus callicarpa* Greene

DESCRIPTION Deciduous clump-forming shrub 6–12 feet (2–3.5 m) high, sometimes large and treelike, with several stems to 2–4 in. (5–10 cm) d.b.h., rarely a small tree to 20 feet (6 m) high and 5 in. (12.5 cm) d.b.h.

LEAVES opposite, compound, pinnate, 5–10 in. (12.5–25 cm) long, with small narrow stipules about 1/8 in. (3 mm) long soon shedding and leaving ring scar on twig, with unpleasant odor. **Leaflets** 5 or 7, paired except at end, short-stalked. **Blades** lanceolate or elliptic, 2–5 in. (5–12.5 cm) long and 1–2 in. (2.5–5 cm) wide, long-pointed at tip and short-pointed and often unequal at base, finely and sharply toothed on edges, thin, above green and nearly hairless, beneath paler and hairy.

TWIGS stout, finely hairy when young, gray, with raised brown dots (lenticels), with rings at nodes. **Buds** paired, large, egg-shaped, 1/4–1/2 in. (6–12 mm) long, gray, covered by several slightly hairy overlapping scales often persistent around twig.

BARK light to dark gray or brown, smoothish, becoming cracked or furrowed into small scaly or shaggy plates.

WOOD soft, whitish. Pith thick, whitish on youngest twigs, becoming deep yellow-orange or brown.

FLOWER CLUSTERS (compound cymes) terminal, erect, longer than broad, 2–4 in. (5–10 cm) long and 1½–2 in. (4–5 cm) wide, with many small whitish flowers with unpleasant odor, turning brown on drying. **Flower** composed of minute 5–toothed calyx, white spreading 5-lobed corolla 3/16–1/4 in. (5–6 mm) across, 5 stamens inserted at base of corolla and alternate with lobes, and pistil with inferior 3-celled ovary with 1 ovule in each cell, short style, and 3 stigmas.

FRUIT many berrylike drupes about 3/16 in. (5 mm) in diameter with calyx persistent at tip, bright red or scarlet, sometimes orange, containing (3) 1–seeded, poisonous nutlets.

PACIFIC RED ELDER

FLOWERING May–July, fruit maturing July–August.

HABITAT Locally common in moist soil, especially in open areas and on recently cutover coastal forests.

NOTE Elders are easily detected by a strong odor when leaves or stems are crushed. The red fruits are classed as not edible, at least when raw, but are sometimes made into wine. Fruits are eaten by some birds, especially robins and thrushes. The "seeds" (nutlets) are reported to be poisonous, causing diarrhea and vomiting. Plants can be grown as ornamentals, but in the interior, the plants only thriving in moist situations.

PACIFIC RED ELDER

PACIFIC RED ELDER

Squashberry
Viburnum edule (Michx.) Raf.

OTHER NAMES mooseberry

SYNONYM *Viburnum pauciflorum* La Pylaie.

DESCRIPTION Deciduous shrub 2–12 feet (0.6–3.5 m) high with several to many stems to 1½ in. (4 cm) d.b.h., sometimes larger and treelike.

LEAVES opposite, with petioles 1/4–3/4 in. (6–20 mm) long, slightly hairy when young, without stipules. **Blades** rounded, thin, mostly shallowly and palmately 3-lobed, 1–4 in. (2.5–10 cm) long and wide, with 3 main veins from rounded base which usually has 2 glands, edges sharply toothed and lobes short-pointed, above dull green and hairless, beneath light green and often hairy, especially on veins.

TWIGS light gray, hairless, stout, with rings at nodes and thick white pith. **Buds** narrowly elliptic, 1/8–1/2 in. (3–12 mm) long, covered by 2 dark red brown, partly united hairless scales, the side buds paired.

BARK gray, smooth.

FLOWER CLUSTERS (cymes) terminal on short lateral twigs bearing only 2 leaves, with persistent stalks 1/2–1 in. (1.2–2.5 cm) long, small, 1/2–1 in. (1.2–2.5 cm) wide, with many or several short-stalked whitish flowers 1/4 in. (6 mm) long and wide. **Flower buds** white or tinged with pink. **Flowers** composed of short 5-toothed calyx, whitish corolla 1/4 in. (6 mm) across the 5 nearly equal spreading lobes, 5 short stamens inserted on corolla and alternate with lobes, and pistil with inferior 3-celled ovary, 1 ovule, and minute stigma.

FRUIT an elliptic red or orange drupe 3/8–1/2 in. (10–12 mm) long, with calyx at tip, sour and edible, containing 1 rounded flat stone 3/16 in. (5 mm) long.

FLOWERING May–July, fruit maturing July–September.

HABITAT Scattered to common, sometimes abundant, in thickets, forest openings, and along streams.

USES The fruits are edible, as the scientific name indicates. They make excellent jelly or juice, especially if picked before mature. Later the flavor may be musty. The flavor of the jelly is

SQUASHBERRY

improved if the juice is mixed with rose hip puree. The over-ripe berries give a musty odor to many areas of interior Alaska in late fall. Wildlife browse the foliage, and numerous birds eat the berries. This species has been recommended for cultivation for its brilliant red autumnal foliage.

ADDITIONAL SPECIES **Highbush-Cranberry** (*Viburnum opulus* L.) is an introduced shrub reported from the Kenai Peninsula. Distinguished from the native *V. edule* by its more deeply 3-lobed leaves, and the inflorescence with an outer margin of sterile white flowers much larger than the fertile inner ones (see photo below).

SQUASHBERRY

HIGHBUSH-CRANBERRY

■ GINSENG FAMILY *Araliaceae*

Mostly tropical trees and shrubs, represented in Alaska by one species of herb and the following spiny shrub. **Leaves** various but often palmately lobed or compound and large, with stipules often forming sheathlike base. **Flowers** small, inconspicuous, greenish, in spreading rounded clusters (umbels), 5-parted, with inferior ovary. **Fruit** a berry; often flattened and 2-celled.

Devil's-club
Oplopanax horridus (Small) Miq.

DESCRIPTION Large deciduous spiny shrub 3–10 feet (1–3 m) high, with few or several thick stems and very few branches. **LEAVES** few, alternate, very large, with long stout spiny hairy petiole 6–12 in. (15–30 cm) long. **Blades** rounded, 6–14 in. (15–35 cm) or more in diameter, thin, palmately 5–9-lobed, lobes sharp pointed and irregularly sharply toothed, heart-shaped at base, with spines along veins, above dull green and hairless, beneath light green and slightly hairy. Stems, petioles, and veins densely covered with many sharp slender yellowish spines or bristly prickles 1/4–3/8 in. (6–10 mm) long.

STEMS thick, 1/2–1 in. (1.2–2.5 cm) in diameter, light brown, with very large white pith. **Buds** large, 1/2 in. (12 mm) long, elliptic, blunt-pointed, brownish tinged, of large, nearly hairless, overlapping scales.

FLOWER CLUSTERS (umbels in a raceme or panicle) terminal, erect, 4–12 in. (10–30 cm) long, nearly stalkless. **Flowers** many, greenish white, 1/4 in. (6 mm) long, fragrant, composed of calyx of 5 teeth, 5 petals 1/8 in. (3 mm) long, 5 alternate stamens longer than petals, and pistil with inferior 2-celled ovary and 2 spreading styles.

FRUITS numerous bright red berries 1/4–3/8 in. (6–10 mm) long, rounded but slightly flattened, with 2 styles at tip, 2-seeded, not edible.

FLOWERING in June, fruits persistent over winter.

HABITAT Common in ravines and openings, moist well-drained soil, characteristic of undergrowth and forming impenetrable thickets in coastal and floodplain forests,

DEVIL'S-CLUB

especially under alder and on good Sitka spruce sites. Plants grow best under partial shade and decline in vigor after clearcutting and exposure to full sunlight. Southeast Alaska north to south-central Alaska, eastern part of Alaska Peninusla, and Kodiak Island. A closely related species or variety is found in Japan.

USES In spite of their spiny nature, the young shoots are browsed by deer and elk in spring and early summer. Alaska Natives sometimes brew tea from the very bitter bark as a tonic or may strip off the thorns and eat the green bark as a tonic. Years ago the stalks were used by Alaska Natives for beating suspected witches to obtain confessions. Even today, old people will nail the devil's-club stalk over their door or window to protect the house from witches, evil influences, and bad luck. Devil's-club is sometimes used as an ornamental in southeast Alaska and southward. It can be planted where seen but not touched, such as in corners and fences. In the fall the foliage provides a prominent splash of yellow.

NOTE The numerous sharp spines are painful and fester when imbedded in the skin, making this shrub dangerous and to be avoided. However, the plants are handsome because of the bright red berries and beautiful mosaic of large leaves arranged to catch the maximum amount of filtered sunlight at the forest edge.

DEVIL'S-CLUB

■ ASTER FAMILY *Asteraceae*

This very large plant family is well represented in Alaska by numerous species of herbs and a single genus with 2 species woody near their base. **Leaves** in this family are various, often toothed or lobed, without stipules. **Flowers** small, crowded and stalkless in heads bordered by greenish scales (bracts), with tubular corolla 5-toothed or strap-shaped (ray flowers), calyx of hairs (pappus) or minute scales or none, and inferior ovary. **Fruit** a dry, 1-seeded achene.

SAGEBRUSH, WORMWOOD *Artemisia*

Sagebrush or wormwood (*Artemisia*) is the only genus of this family with woody plants in Alaska. Of about 20 Alaskan species, only 2, fringed sagebrush and Alaska sagebrush, can be considered as shrubs, though they are primarily herbaceous and woody only near base. Both grow on the driest, warmest sites in central and northern Alaska: steep, rocky, south-facing river bluffs.

KEY TO ALASKA *ARTEMISIA*

1   Basal leaves 2 to 3 times divided into linear segments 1/32 in. (1 mm) wide, leaf blade 1/4–1/2 in. (6–12 mm) long   FRINGED SAGEBRUSH (*Artemisia frigida*)
1   Basal leaves 2–3 times divided into spatula-shaped (spatulate or oblanceolate) segments 1/16–1/8 in. (2–3 mm) wide, leaf blade 1–2 in. (2.5–5 cm) long
                                              ALASKA SAGEBRUSH (*Artemisia alaskana*)

Alaska wormwood

Artemisia alaskana Rydb.

SYNONYM *Artemisia krushiana* Bess, *Artemisia tyrellii* Rydb.
DESCRIPTION Silvery spreading shrub, 18–24 in. (45–60 cm) high, much branched from woody base, fragrant.
BASAL LEAVES 1–2 in. (2.5–5 cm) long, divided into 3–5 segments, each again divided into spatula-shaped (spatulate), oblong, or linear segments 1/16 in. (2 mm) wide; **stem leaves** becoming less divided so that upper leaves may be undivided and linear-shaped; present year's leaves densely silky hairy, past year's basal leaves usually persisting, gray brown in color.

STEMS herbaceous, silvery from dense hairs, dying back each winter to a few short basal woody stems; older stems brown and covered with old dead gray leaves.

FLOWERS in compact heads 1/4–5/16 in. (6–8 mm) in diameter, often nodding, on a branched, narrow, erect leafy twig (raceme), yellow, and lacking ray flowers, the bracts with dense silvery hairs.

FRUITS many tiny hairless seeds (achenes).

FLOWERING in July and August, seeds maturing August and September.

HABITAT Common along the river bluffs of central Alaska and occasionally along rivers north of the Brooks Range.

NOTE Sometimes treated as *Artemisia krushiana* subsp. *alaskana* (Rydb.) D.F. Murray & Elven; *krushiana* also spelled as *krusiana*.

ALASKA
WORMWOOD

Fringed Sagebrush
Artemisia frigida Willd.

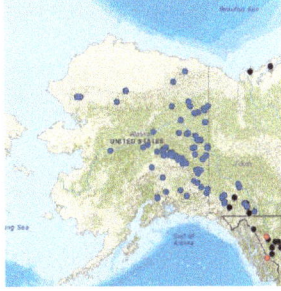

OTHER NAME prairie sagewort

DESCRIPTION Shrubby spreading perennial, much branched from woody base, 12–18 in. (30–45 cm) high, fragrant, and silvery in appearance.

LEAVES densely crowded at base and along stem, small and divided 2 or 3 times into linear segments less than 1/32 in. (1 mm) wide, total length of blade 3/16–1/2 in. (5–12 mm), densely silky hairy throughout.

STEMS of current year herbaceous, silvery from dense white hairs, dying back each winter to a few short woody stems, older woody stems covered with dead gray leaves, silvery in some parts but becoming brown with age.

FLOWERS in small compact heads about 1/8 in. (3–4 mm) in diameter, on a narrow erect leafy branch (raceme), yellow without ray flowers, the underlying bracts with dense silvery hairs.

FRUITS many tiny hairless seeds (achenes).

FLOWERING in July–August, seeds maturing August–September.

HABITAT Common on sunny, south-facing, well drained river bluffs in central Alaska, too dry or unstable for trees. In the summer it may be confused with other herbaceous species of *Artemisia,* which have much larger, less dissected leaves. Known along river bluffs of major rivers of central Alaska, including Matanuska, Copper, Kuskokwim, Tanana, and Yukon.

FRINGED SAGEBRUSH

FRINGED SAGEBRUSH

■ BIRCH FAMILY *Betulaceae*

The birch family (Betulaceae) is represented in Alaska by two genera, birch (*Betula*) and alder (*Alnus*), and 8 species (also intergrading varieties and hybrids). Distinguishing characters are:

• **Leaves** borne singly (alternate), broad, margins sharply and usually doubly toothed with teeth of 2 sizes, and in alders often slightly wavy lobed;

• **flower clusters** (catkins) composed of an axis bearing many minute greenish flowers 2–3 above a scale, in early spring before the leaves, from buds partly formed the preceding summer;

• **flowers** with minute calyx, of 2 kinds on the same plant (monoecious);

• **male flowers** with pollen in long, narrow catkins at end of twig and female flowers in short catkins on sides of twig; and

• **fruits** conelike, 1/2–2 in. (1.2–5 cm) long, of many nutlets ("seeds") and scales.

The **tree birches** of Alaska are easily recognized by their smooth, thin, white, pinkish, coppery brown, or purplish brown bark, which peels off in papery strips; the soft conelike fruits shed, leaving a slender axis. **Alders** generally have smooth gray bark, which is not papery, and usually have at all seasons some old dead, hard, blackish or dark grown conelike fruits remaining on the twigs. **Birch twigs** commonly have raised gland dots and have winter buds not stalked, composed of overlapping scales. **Alder twigs** lack glands and have usually stalked winter buds with 3 exposed scales usually meeting at their edges or overlapping.

ALDER *Alnus*

Alaska has three alder species, two of which reach tree size. Alders are easily recognized by their smooth gray bark with horizontal lines (lenticels) and the clusters of 3–9 slender-stalked old dead, hard, blackish or dark brown conelike elliptic fruits generally present. **Male flowers** in narrow catkins, 3 above a scale, composed of 4 sepals and usually 4 stamens. **Female catkins** short, about 1/2 in. (12 mm) long; flowers in group of 2 above a scale, composed of ovary and 2 styles. Alder roots, like those of legumes, often have root nodules, or swellings that fix nitrogen from the air and enrich the soil.

THINLEAF ALDER

KEY TO ALASKA ALDERS

1 Leaves yellow green above, shiny on both sides and especially beneath, sticky when young, edges with relatively long-pointed teeth; stalks about as long as conelike fruits; nutlets with 2 broad wings; winter buds of overlapping scales   SITKA ALDER (*Alnus sinuata*)

1 Leaves dark green above, dull, not sticky when young, edges with short-pointed teeth; stalks shorter than conelike fruits; nutlets with 2 narrow wings or none; winter buds of 3 exposed scales meeting at edges   2

2 Leaves thick with edges curled under slightly, with rusty hairs along veins beneath; conelike fruits 1/2–1 in. (12–25 mm) long; nutlets with 2 narrow wings   RED ALDER (*Alnus rubra*)

2 Leaves thin with edges flat, finely hairy or nearly hairless beneath; conelike fruits 3/8–5/8 in. (10–15 mm) long; nutlets almost wingless   THINLEAF ALDER (*Alnus incana*)

Thinleaf Alder

Alnus incana (L.) Moench

OTHER NAME speckled alder

SYNONYM *Alnus tenuifolia* Nutt.; Alaska plants are subsp. *tenuifolia* (Nutt.) Breitung

DESCRIPTION Deciduous large shrub or small tree 15–30 feet (4.5–9 m) high, commonly forming clumps, with trunks to 8 in. (20 cm) in diameter.

LEAF BLADES ovate or elliptic, 2–6 in. (5–15 cm) long, $1\frac{1}{4}$–$2\frac{1}{2}$ in. (3–6 cm) wide, short-pointed, rounded at base, shallowly wavy lobed and doubly toothed with both large and small teeth, thin, dark green and becoming hairless above, beneath pale green and hairy or nearly hairless; petioles 1/4–1 in. (6–25 mm) long.

TWIGS reddish and hairy when young, becoming gray.

BARK gray to dark gray, smooth, becoming reddish gray, thin and scaly.

WOOD light brown.

MALE FLOWERS in narrow catkins $1\frac{1}{2}$–3 in. (4–7.5 cm) long.

FRUITS on short stalks less than 1/4 in. (6 mm) long, conelike, 3/8–5/8 in. (1–1.5 cm) long; **nutlets** elliptic, almost wingless.

FLOWERING in May–June.

HABITAT Commonly forms thickets (often with the larger willows) along streams in central and southern Alaska.

USES Large trunks have been cut for poles. The wood is used for smoking salmon.

THINLEAF ALDER

THINLEAF ALDER

SPECKLED ALDER

Red Alder
Alnus rubra Bong.

OTHER NAME western alder

SYNONYM *Alnus oregona* Nutt.

DESCRIPTION Small to medium-sized deciduous tree 20–40 feet (6–12 m) tall, with straight trunk 4–16 in. (10–40 cm) in diameter.

LEAF BLADES ovate or elliptic, 3–5 in. (7.5–12.5 cm) long, 1¾ –3 in. (4.5–7.5 cm) wide, short pointed at both ends, shallowly wavy lobed and doubly toothed with both large and small teeth, thick, edges curled under slightly, dark green and nearly hairless above, beneath pale with rusty hairs along veins; petioles 1/4–3/4 in. (6–20 mm) long.

TWIGS hairy when young, becoming dark red with light dots. **Winter buds** stalked, to 3/8 in. (1 cm) long, dark red.

BARK gray, splotched with white, smooth or becoming slightly scaly, thin.

WOOD nearly white when freshly cut, soon turning to light reddish brown, fine-textured, moderately lightweight, soft.

MALE FLOWERS in narrow catkins 3–6 in. (7.5–15 cm) long.

FRUITS on short stalks 1/4–1/2 in. (6–12 mm) conelike, 1/2–1 in. (12–25 mm) long; **nutlets** elliptic, with 2 narrow wings.

FLOWERING in April-May.

HABITAT Red alder is common throughout southeast Alaska on stream bottoms with rich, rocky, moist soils and along beaches where

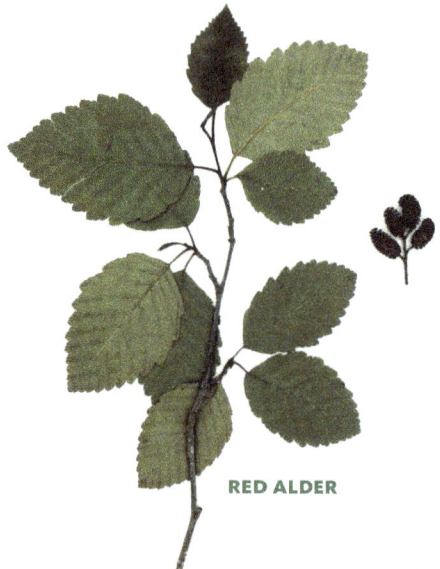

RED ALDER

creeks enter the sea. On landslides it forms almost impenetrable thickets, often with Sitka alder. Red alder is a pioneer species on mineral soil, thriving on moist sites. It is common below 1,000 feet elevation and absent at higher elevations, where Sitka alder is frequent. Being larger, red alder is more competitive and requires more time for overtopping. Both species come in along roadsides and where ground is disturbed after logging. They are a problem in road maintenance, requiring continual clearing of shoulders and side slopes. Seeds of both species are produced within five years, and being tiny are blown great distances.

RED ALDER - FEMALE CONES

RED ALDER

Sitka Alder
Alnus viridis (Chaix) DC.

SYNONYM *Alnus crispa* (Ait.) Pursh subsp. *sinuata* (Reg.) Hultén, *Alnus sitchensis* (Reg.) Sarg., *Alnus sinuata* (Regel) Rydb.

DESCRIPTION Deciduous shrub 5–15 feet (1.5–4.5 m) high or a small tree to 30 feet (9 m) tall and 8 in. (20 cm) in trunk diameter, often with multiple trunks.

LEAF BLADES ovate, 2½–5 in. (6–12.5 cm) long, 1½–3 in. (4–7.5 cm) wide, short-pointed, rounded at base, shallowly wavy lobed and doubly toothed with long-pointed teeth of 2 sizes, sticky when young, speckled yellow green and shiny above, beneath lighter, shiny, and hairless or nearly so; petioles 1/2–3/4 in. (12–20 mm) long.

TWIGS sticky, finely hairy, and orange brown when young, becoming light gray. **Winter buds** short-stalked to stalkless on young twigs, to 1/2 in. (12 mm) long, of overlapping scales.

BARK gray to light gray, smooth and thin.

MALE FLOWERS in narrow catkins 3–5 in. (7.5–12.5 cm) long.

FRUITS 1/2–3/4 in. (12–20 mm) long, on long slender spreading stalks 3/8–3/4 in. (10–20 mm) long, **nutlets** elliptic, with 2 broad wings.

FLOWERING in May–June.

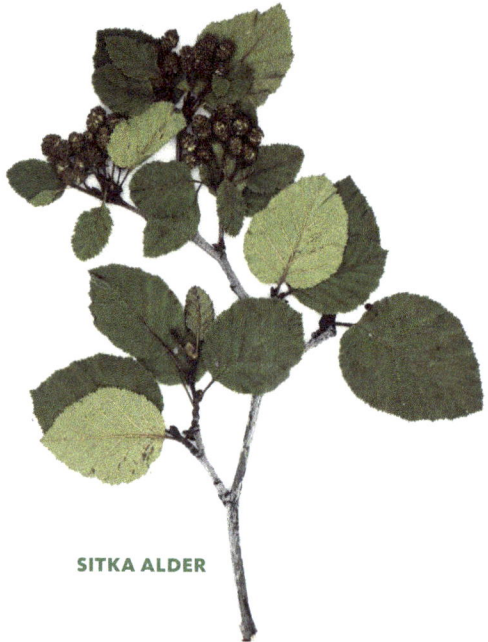

SITKA ALDER

HABITAT common to abundant, with many stems, forming thickets in marshes, along streams, on landslides and gravelly slopes, and in clearings, from sea level to the alpine zone above the timberline.

NOTE This pioneer species follows disturbances such as landslides, logging, or glacial retreat. It requires mineral soil seedbed and develops rapidly on moist sites but grows on soils too sterile for other trees. Sitka spruce often becomes established at the same time. Alder acts as a nurse tree, improving soil conditions, and adding organic matter and nitrogen. It thrives with overhead light but is intolerant of shade and disappears from the stand when overtopped by Sitka spruce. Being smaller and hence more quickly overtopped, Sitka alder is probably not such a serious competitor as red alder on logged areas.

USES The wood produces good fuel and is used for smoking fish. Alder twigs and buds make up an important part of the winter food of the white-tailed ptarmigan. In the fall and winter the "seeds" (nutlets) are eaten by many songbirds.

MALE FLOWERS

SITKA ALDER

SITKA ALDER - FEMALE CONES

BIRCH *Betula*

Alaska has two species of dwarf, shrubby birches, both widely distributed, and three kinds of tree birches. Tree species are variable and intergrade and hybridize wherever their ranges meet. The dwarf birches have round, rounded-toothed leaves less than 3/4 in. (2 cm) long, while the tree birches have larger, ovate leaves 1½–3½ in. (4–9 cm) long.

The tree birches of Alaska were formerly treated as three geographical varieties of a single transcontinental species, paper birch (*Betula papyrifera* Marsh.). Most modern treatments now separate them as follows:

• **Paper birch** (*B. papyrifera*) has leaves mostly rounded at base and usually reddish brown bark.

• **Alaska paper birch** (*B. neoalaskana*) has rather long-pointed leaves usually wedge-shaped at base and usually white bark in age (or reddish brown when young or in dense stands).

• **Kenai birch** (*B. kenaica*) has relatively thick, usually short-pointed leaves and usually dark brown or gray bark.

KEY TO ALASKA BIRCHES

1 Leaf blades rounded or elliptic, thick, less than 1¼ in. (3 cm) long, rounded teeth on edges; shrubs or sometimes small trees with smooth bark not peeling **2**

1 Leaf blades ovate, 1½–3½ in. (4–9 cm) long, mostly thin, with pointed teeth on edges; trees with thin papery bark, peeling off **4**

2 Leaf blades less than 3/4 in. (2 cm) long; low shrubs less than 5 ft (1.5 m) high **3**

2 Leaf blades 1–1¼ in. (2.5–3 cm) long; large shrubs or trees becoming more than 10 ft (3 m) high **HYBRID BIRCHES** (*Betula* hybrids)

3 Leaf blades often broader than long, 3/16–1/2 in. (5–12 mm) long, straight or notched at base **DWARF ARCTIC BIRCH** (*Betula nana*)

3 Leaf blades longer than broad, mostly 3/8–3/4 in. (10–20 mm) long, wedge-shaped at base **RESIN BIRCH** (*Betula glandulosa*)

4 Leaves long-pointed, usually wedge-shaped at base; bark usually white in age (or reddish brown when young or in dense stands); interior Alaska **ALASKA PAPER BIRCH** (*Betula neoalaskana*)

4 Leaves mostly short-pointed; bark brown or pinkish; southern and southeast Alaska **5**

5 Leaves thin, mostly rounded at base; bark usually reddish brown **PAPER BIRCH** (*Betula papyrifera*)

5 Leaves thick, wedge-shaped or rounded at base, with white hairs on toothed edges; bark usually dark brown or gray; southern and southern interior Alaska **KENAI BIRCH** (*Betula kenaica*)

Resin Birch
Betula glandulosa Michx.

OTHER NAMES shrub birch, glandular scrub birch, bog birch, ground birch, dwarf birch

DESCRIPTION Deciduous shrub mostly low and spreading to erect, 1–5 feet (0.3–1.5 m) high or taller, forming clumps. **LEAVES** with short hairy petioles 3/16–1/4 in. (5–6 mm) long. **Blades** elliptic to broadly obovate, mostly 3/8–3/4 in. (1–2 cm) long, rounded but longer than broad, rounded at tip, finely wavy-toothed except near wedge-shaped base, thick and leathery, often with glandular dots on both surfaces (visible only with high magnification); above shiny dark green and usually hairless, beneath yellow green and often finely hairy.

TWIGS often finely hairy when young, densely resinous with warty glands, with a gray layer of wax.

BARK reddish brown, becoming dark gray, smooth, not peeling.

MALE FLOWER CLUSTERS (catkins) several near base of twigs, 1/2–1 in. (12–25 mm) long, 3/16–1/4 in. (5–6 mm) wide, of light brown scales and numerous stamens.

FEMALE FLOWER CLUSTERS several to many on older twigs 1/4–1/2 in. (6–12 mm) long, 1/16 in. (2 mm) wide, greenish.

FRUITS conelike, 3/8–1 in. (10–25 mm) long, 1/8–1/4 in. (3–6 mm) wide, mostly erect, with many 3-lobed bracts or scales with resinous dot or hump on back. **Nutlets** elliptic,

RESIN BIRCH

RESIN BIRCH - YOUNG FEMALE CONES

RESIN BIRCH

flattened, reddish brown, more than 1/16 in. (2 mm) long, with 2 very narrow wings narrowest at base.

FLOWERING May–June, fruits maturing July–August, persistent in winter.

HABITAT Moist soil, especially in muskegs or boggy areas, hummocks on tundra, and borders of lakes and streams. Forming extensive thickets at tree-line in the Alaska and Brooks Ranges.

USES The leaves and young twigs are browsed by caribou and reindeer. In winter, the buds and twigs are clipped by ptarmigan.

ADDITIONAL SPECIES Many birches in Alaska exhibit leaf and bark characters intermediate between the tree and shrub birches described here, and are considered to be of hybrid origin. Perhaps the most common is **water birch**, believed to be a hybrid between **resin birch** and **Alaska paper birch** (*Betula glandulosa* X *B. neoalaskana*; also often treated as *Betula occidentalis* Hook.), and found throughout interior Alaska. These plants are common near tree-line, where birch trees below meet a band of resin birch shrubs above. Plants are large, spreading, and clump-forming, 10–12 feet (3–3.7 m) high, with many stems 1 in. (2.5 cm) in diameter, sometimes becoming a small tree 15–20 feet (4.5–6 m) high and 3–6 in. (7.5–15 cm) in diameter. Leaves have slender petioles 1/4–3/8 in. (6–10 mm) long; leaf blades are elliptic to diamond-shaped, 1–1¼ in. (2.5–3 cm) long, 3/4–1¼ in. (2–3 cm) wide, short-pointed or rounded at both ends, with rounded teeth on edges, thick, becoming hairless. Twigs are often densely covered with glandular dots. Bark is reddish black, smooth and not peeling. Fruits are conelike 3/4 in. (2 cm) long.

WATER BIRCH

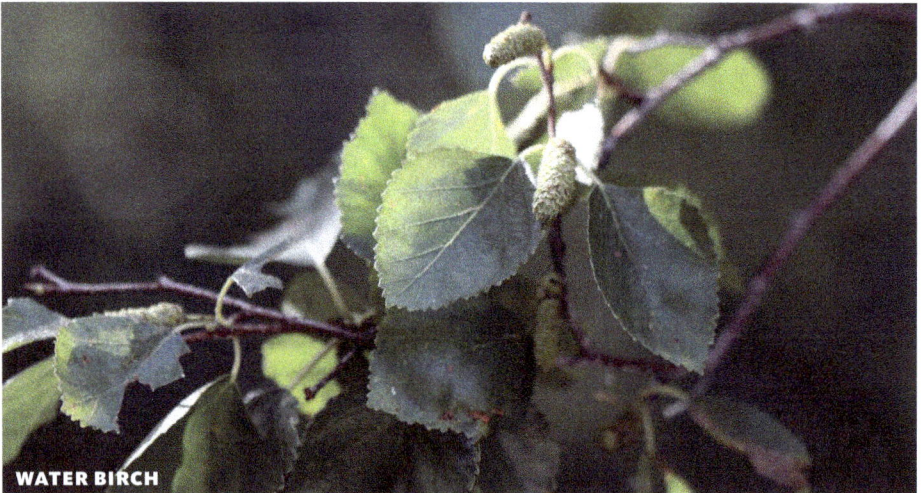

WATER BIRCH

Kenai Birch
Betula kenaica W.H. Evans

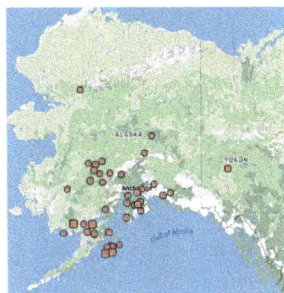

OTHER NAMES Kenai paper birch, black birch, red birch
SYNONYM *Betula papyrifera* var. *kenaica* (W.H. Evans) Henry
DESCRIPTION Small to medium-sized tree 20–80 feet (6–24 m) high and 4–12 in. (10–30 cm) in trunk diameter, rarely 18 in. (46 cm).
LEAF BLADES ovate or nearly triangular, 1½–2 in. (4–5 cm) long, 1–1¾ in. (2.5–4.5 cm) wide, relatively thick, usually short-pointed, broadly wedge-shaped or rounded at base, margin coarsely and often doubly toothed with white hairs, dull dark green and often slightly hairy above, beneath pale yellow green and dotted with glands and hairy on veins; petioles usually hairless.
TWIGS reddish brown hairy, and often with resin dots when young, becoming blackish and hairless.
BARK usually dark brown, often blackish or reddish brown, sometimes becoming pinkish or grayish white, papery.
MALE FLOWER CATKINS short, about 1 in. (2.5 cm) long, narrow, dark brown.
FRUITS conelike, about 1 in. (2.5 cm) long, erect or spreading. **Nutlets** with wings slightly narrower than body; bracts, with lobes of about equal length, rounded at tip, side bracts slightly diamond-shaped.
HABITAT Kenai birch, named from Kenai Peninsula and known only from Alaska and the Yukon Territory, is found in the southern part of the spruce-birch interior forests (but not southeast Alaska).
NOTE *Betula kenaica* differs from *B. papyrifera* primarily in its smaller size and in its smaller, blunter-tipped, more coarsely and regularly serrate leaves.

KENAI BIRCH

KENAI BIRCH

Dwarf Arctic Birch
Betula nana L.

OTHER NAMES dwarf birch, dwarf alpine birch
SYNONYM *Betula glandulosa* var. *sibirica* (Ledeb.) Blake, *Betula pumila* L.
DESCRIPTION Low spreading deciduous shrub commonly 1/2–3 feet (1.5–9 dm) high.
LEAVES alternate, almost stalkless, with slender petioles 1/16 in. (2 mm) long. **Blades** round or kidney-shaped, often broader than long, 3/16–1/2 in. (5–12 mm) long, 3/16–5/8 in. (5–16 mm) wide, rounded at tip, finely wavy toothed to straight or notched base, thick, hairless, above green, beneath pale green, turning copper-red in autumn.
TWIGS slightly resinous and slightly hairy, with a few tiny, warty glands.
MALE FLOWER CLUSTERS 3/8–1 in. (1–2.5 cm) long, with brown scales.
FEMALE FLOWER CLUSTERS 1/4–3/8 in. (6–10 mm) long, green.
FRUITS conelike, elliptic, 3/16–1/2 in. (5–12 mm) long, 3/16–1/4 in. (5–6 mm) wide, light brown, with many 3-lobed bracts or scales without resinous dot or hump on back. **Nutlets** many, elliptic, with 2 narrow wings of equal width from base to tip.
FLOWERING June, fruits maturing July–August.
HABITAT Moist soil, muskegs or bogs, rocky alpine slopes, and hummocks on tundra. Widespread nearly throughout Alaska over the coasts and in mountains of interior from northern part of southeast Alaska to western end of Alaska Peninsula and Bering Sea, north to Arctic Coast.

DWARF ARCTIC BIRCH

DWARF ARCTIC BIRCH

Alaska Paper Birch
Betula neoalaskana Sarg.

OTHER NAMES Alaska white birch, Alaska birch, canoe birch, paper birch, white birch

SYNONYMS *Betula papyrifera* var. *humilis* (Reg.) Fern. & Raup, *Betula papyrifera* var. *neoalaskana* (Sarg.) Raup, *Betula alaskana* Sarg., *Betula pendula* subsp. *mandshurica* (Regel) Ashburner & McAll.

DESCRIPTION Small to medium-sized tree 20–80 feet (6–24 m) high and 4–24 in. (10–61 cm) in trunk diameter.

LEAF BLADES ovate, 1½–3 in. (4–7.5 cm) long, 1–2 in. (2.5–5 cm) wide, rather long-pointed, sharply to broadly wedge-shaped at base, coarsely toothed, dark green or yellow-green and hairless above, beneath pale yellow-green, dotted with glands and usually with angles of lower veins hairy; petioles becoming hairless.

TWIGS with many raised resinous dots (glands).

BARK white, or pinkish white, sometimes grayish white or yellowish white, papery.

MALE FLOWER CATKIN short, 1–1½ in. (2.5–4 cm) long, thick, greenish brown.

FRUITS conelike, 1–1 3/8 in. (2.5–3.5 cm) long, hairless, hanging down or spreading. **Nutlets** with wings broader than body; bracts with middle lobe usually longer than the blunt, diamond-shaped lateral lobes, hairy on edges.

HABITAT Common in spruce-birch forests throughout most of interior Alaska but not in southeast; a common birch through the interior up to tree-line. It is best developed on warm slopes with moist, porous soils, but is also common on cold north slopes and poorly drained lowlands following fires. In general, the tree-sized birches of Alaska grow in a mix with white or black spruce, the spruce eventually replacing birch in the successional sequence following fire.

At Cook Inlet there are important birch forests. Here, Alaska paper birch has its best development on the rolling benchlands and lower foothill slopes up to an altitude of about 800 feet.

NOTE Easily confused with *Betula papyrifera,* but *B. neoalaskana* twigs are covered with small, raised, resin glands.

ALASKA PAPER BIRCH

ALASKA PAPER BIRCH

Paper Birch
Betula papyrifera Marsh.

OTHER NAMES white birch, canoe birch

DESCRIPTION Small to medium-sized deciduous tree usually 20–60 feet (6–18 m) high and 4–12 in. (10–30 cm) in trunk diameter, becoming 80 feet (24 m) tall and 24 in. (60 cm) in diameter.

LEAVES with slender petioles 1/2–1 in. (1.2–2.5 cm) long. **Leaf blades** ovate, 1½–3½ in. (4–9 cm) long, 1–2½ in. (2.5–6 cm) wide, long-pointed or short-pointed at tip, wedge-shaped or rounded at base, coarsely and usually doubly toothed, mostly dull dark green and hairless above, light yellow-green and hairless or slightly hairy beneath.

TWIGS slender, hairless, reddish brown with many small whitish dots, with short side twigs (spur shoots) covered by many raised half-rounded leaf-scars, becoming reddish black. **Winter buds** conic, 1/4 in. (6 mm) long, long-pointed, dark brown, slightly resinous, covered by 3 overlapping scales.

BARK smooth, with long horizontal lines (lenticels), thin, separating into papery strips and peeling off, from white to pinkish, coppery brown, or purplish brown in the different varieties; inner bark orange.

WOOD of wide white sapwood and light reddish brown heartwood, fine-textured, moderately hard, and moderately heavy (the densest of Alaska commercial woods).

FLOWERS male and female on same twig, tiny, in groups of 3 above a scale (bract). **Male flowers** in narrow catkins partly developed the preceding summer, 1–4 in. (2.5–10 cm) long composed of calyx and 2 stamens; **female flowers** in shorter clusters 3/8–1 in. (1–2.5 cm) long, composed of ovary and 2 styles.

PAPER BIRCH

FRUITS conelike, cylindric, 1–2 in. (2.5–5 cm) long and 3/8 in. (1 cm) wide, slender-stalked and hanging down. **Nutlets** ("seeds") many, 1/16 in. (1.5 mm) long, brown, with 2 broad wings.

FLOWERING in May–June, before the leaves, fruit shedding gradually into winter.

HABITAT Paper birch is a characteristic species of the interior forests of Alaska, designated as spruce-birch forests, and is associated with white spruce and aspen. In the upper Cook Inlet area, extensive paper birch forests occupy the rolling benchland above the bottoms, and extend up the slopes of the foothills to about 800 feet (244 m). Growth is moderate to fast. On more favorable sites, trees 80 to 100 years old attain a height of 60–70 feet (18–21 m) and a trunk diameter of 12–14 in. (30–35 cm) Average diameter is 8–10 in. (20–25 cm) and maximum about 29 in. (73 cm). Maximum age recorded is about 230 years.

USES Near cities and villages in Interior Alaska, paper birch has been used primarily for fuel, mainly fireplace wood. It has served locally for mine props. A small amount of lumber is cut and marketed locally in Interior Alaska. However, attempts to develop export markets have not yet been successful because of high costs and transportation problems. The wood has been made into cabinets and wall paneling.

The wood of paper birch varieties growing in Alaska is suitable for pulping and paper-making by several processes. It is satisfactory also for furniture, cabinetmaking, veneer and plywood, handles, boxes and crates, clothes pins, spools, and bobbins. Other uses of paper birch southward are turned and carved articles, toothpicks, and toys. The wood works easily and takes finishes and stains satisfactorily. The uniformity of grain is a distinct advantage in the manufacture of veneers and plywoods.

Native Americans made canoes and various small articles from the smooth thin bark. Because of its durability and ease of working, bark was used as sheeting under sod on cabin roofs. With their attractive bark, birch trees are planted as ornamentals.

NOTE Paper birch is one of the most widespread tree species in northern North America and is composed of a number of intergrading geographical varieties or closely related species.

PAPER BIRCH - BARK

PAPER BIRCH

■ HONEYSUCKLE FAMILY *Caprifoliaceae*

Deciduous (ours) or evergreen shrubs, sometimes small trees, woody vines, and herbs. Leaves opposite, simple or pinnately compound, without stipules (present in *Sambucus*). Flowers mostly small, regular or irregular, composed of calyx of 4–5 teeth, tubular corolla with 4–5 lobes, 4–5 stamens inserted on tube and alternate with lobes, and pistil with inferior ovary of 2–5 cells and usually 1 ovule in each and 1 style or none. Fruit mostly a berry or berrylike drupe. Two genera in Alaska, each with a single native species.

HONEYSUCKLE *Lonicera*

Bearberry Honeysuckle

Lonicera involucrata (Richards.) Banks ex Spreng.

OTHER NAMES honeysuckle, black twinberry

DESCRIPTION Deciduous shrub 3–10 feet (1–3 m) high.

LEAVES opposite, with petioles less than 1/4 in. (6 mm) long, without stipules. **Blades** elliptic, 2–5 in. (5–12.5 cm) long, 1–3 in. (2.5–7.5 cm) wide, long-pointed or short-pointed at both ends, edges hairy and not toothed, above dull green and hairless or nearly so, beneath pale green and hairy on veins.

TWIGS 4-angled when young, hairless, ringed at nodes.

BARK becoming gray and shreddy.

FLOWERS paired above 4 leaflike green or purple bracts on stalk 1–2 in. (2.5–5 cm) long at base of leaves, 1/2–5/8 in. (12–15 mm) long, composed of short tubular calyx, yellow funnel-shaped corolla swollen on one side at base and with 5 nearly equal short lobes, 5 glandular hairy stamens inserted within tube, and pistil with inferior 3-celled ovary, many ovules, and slender style.

FRUITS paired above 4 dark red bracts, 3-celled, few-seeded black berries, round, 3/8 in. (10 mm) in diameter.

HABITAT Rare and local in wet soil; southeast Alaska.

NOTE The bitter fruits are said to be poisonous.

ADDITIONAL SPECIES Twinsisters (*Lonicera tatarica* L.), is introduced and reported from Fairbanks, where likely an uncommon escape from plantings.

BEARBERRY HONEYSUCKLE

BEARBERRY HONEYSUCKLE

BEARBERRY HONEYSUCKLE

TWINSISTERS

TWINSISTERS

Common Snowberry
Symphoricarpos albus (L.) Blake

SYNONYM *Symphoricarpos rivularis* Suksd.

DESCRIPTION Deciduous, much branched shrub 1–4 feet (3–12 dm) high sometimes taller.

LEAVES opposite, with slender petioles about 1/8 in. (3 mm) long, without stipules. **Blades** elliptic to ovate, 5/8–1½ in. (1.5–4 cm) long, 1/2–1 in. (1.2–2.5 cm) wide, blunt at both ends, on vigorous twigs larger and often with a few irregular teeth or lobes, thin, above dark green and hairless or nearly so, beneath often whitish green and hairy.

TWIGS slender, reddish brown, hairless or minutely hairy, ringed at nodes, older twigs gray with shreddy bark. Buds 1/16 in. (1.5 mm) long, scaly.

FLOWERS mostly few in short clusters (racemes) at ends of twigs or also at bases of upper leaves, about 1/4 in. (6 mm) long, pink, composed of 5-toothed calyx, pink tubular bell-shaped corolla 1/4 in. (6 mm) long and nearly as wide, 5-lobed, hairy within, with 5 stamens inserted in tube alternate with lobes, and pistil with elliptic inferior 4-celled ovary with 2 ovules and short hairless style.

FRUITS 1 to few, round white berrylike drupes 1/4–5/8 in. (6–15 mm) long, with calyx at tip, containing 2 light brown nutlets or stones 3/16 in. (5 mm) long. Collected in flower in July.

HABITAT Local in southeast Alaska. Cultivated elsewhere as an ornamental.

COMMON SNOWBERRY

COMMON SNOWBERRY

■ DOGWOOD FAMILY *Cornaceae*

Shrubs and trees, represented in Alaska by one species of shrub and three low herbs. **Leaves** in the Alaska species paired, elliptic, without teeth or lobes, with long curved side veins, without stipules. **Flowers** small but often crowded and bordered by showy petallike bracts, 4- or 5-parted, with inferior ovary. **Fruit** a drupe.

Red-Osier Dogwood
Cornus alba L.

SYNONYM *Cornus stolonifera* Michx.
DESCRIPTION Deciduous shrub 3–12 feet (1–3.5 m) high, with several stems, reported to 15 feet (4.5 m) high and treelike. **LEAVES** paired (opposite), with hairy petioles 1/4–1/2 in. (6–12 mm) long, without stipules. **Blades** elliptic to ovate, $1\frac{1}{2}$–$3\frac{1}{2}$ in. (4–9 cm) long and 5/8–2 in. (1.5–5 cm) broad, short or long-pointed at tip, short-pointed or rounded at base, edges not toothed, 5–7 long curved, sunken veins on each side of midrib, dull green and nearly hairless above, finely hairy and whitish green beneath.

TWIGS dark red, mostly finely hairy when young, with rings at nodes, whitish dots (lenticels), and large white pith.

BARK gray, smooth to slightly furrowed into flat thick plates.

FLOWER CLUSTERS (cymes) terminal, flat, $1\frac{1}{4}$–$2\frac{1}{4}$ in. (3–5.5 cm) across, the branches persistent in winter. **Flowers** many, crowded, short-stalked, about 1/4 in. (6 mm) long and broad, finely hairy, composed of calyx of 4 minute sepals united at base, 4 white petals 1/8 in. (3 mm) long, 4 alternate stamens, and pistil with inferior 2-celled ovary and short style.

FRUIT (drupe) round, 1/4–3/8 in. (6–10 mm) in diameter, whitish or light blue, with 1 nutlet 3/16 in. (5 mm) long.

FLOWERING June–July, fruits maturing July–September.

HABITAT Common in moist soil in clearings and in open understory of forests, especially on floodplains of major rivers.

USES Often used as ornamentals and easily propagated by stem cuttings. Young twigs are a preferred browse of moose during fall and winter. The lower branches root at tip (as indicated by the former scientific name), at least in some parts of its broad range.

RED-OSIER DOGWOOD

RED-OSIER DOGWOOD

RED-OSIER DOGWOOD

ADDITIONAL SPECIES

Bunchberry (*Cornus canadensis* L.) has erect stems 4–8 in. (10–20 cm) high. **Leaves** 1 pair small and at summit 4–6 large leaves (whorled), short-stalked, elliptic, 1½–2½ in. (4–6 cm) long and ¾–1½ in. (2–4 cm) wide. **Flowers** many, minute, whitish or yellowish, in a head ¾–1¼ in. (2–3 cm) across, above 4 white petal-like bracts. **Fruit** a cluster of 10 or fewer orange-red round drupes 5/16 in. (8 mm) in diameter. Very common in Alaska except extreme north and to Unalaska Island in Aleutian Islands. Forming ground cover in interior spruce forests. In southeast Alaska used in all seasons as browse by blacktail deer. Propagated as an ornamental ground cover in interior Alaska for its showy flowers, fruits, and fall coloring. It is reported that the berries are sometimes used for jelly and pies.

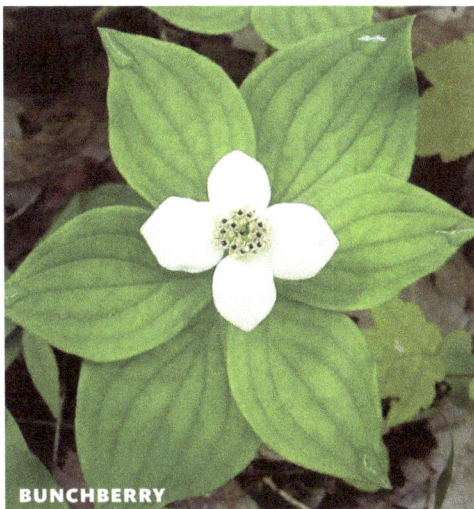

BUNCHBERRY

Dwarf bog bunchberry (*Cornus suecica* L.) has erect stems 2–8 in. (5–20 cm) high, bearing 2–8 paired **leaves** (opposite), ¾–1¼ in. (2–3 cm) long and 5/8–1 in. (1.5–2.5 cm) wide, stalkless, lanceolate to elliptic. **Flowers** many, minute, dark purple, in a head ¾–1 in. (2–2.5 cm) across the 4 white petal-like bracts. **Fruit** a cluster of 3–10 rose-red round drupes about 5/16 in. (8 mm) in diameter.

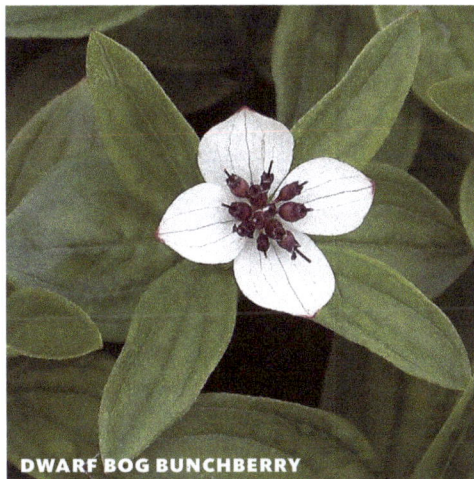

DWARF BOG BUNCHBERRY

Western cordilleran bunchberry (*Cornus unalaschkensis* Ledeb.) has erect stems, 2–10 in. (5–25 cm) tall, minutely bristly with 2-pronged hairs. **Stem leaves** 1–2 pairs below the terminal whorl of 4–6 stalkless leaves, these ovate to elliptic, with lateral veins arising from the base of the leaf (or nearly so); sparsely hairy above, glabrous and paler below. **Flowers** subtended by 4 white, yellowish, pinkish or purplish bracts; petals and sepals dark purplish. **Fruit** a fleshy red drupe, 1/4–1/2 in. (6–12 mm) long.

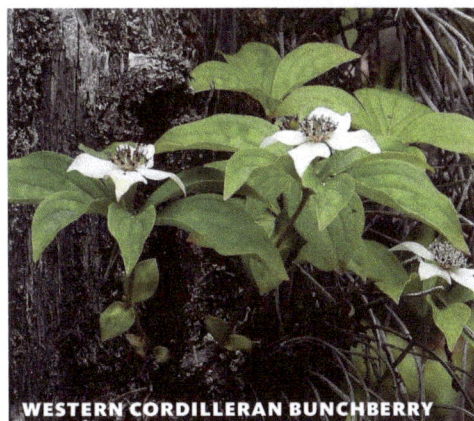

WESTERN CORDILLERAN BUNCHBERRY

■ DIAPENSIA FAMILY *Diapensiaceae*

Evergreen low shrubs with crowded or alternate **leaves** without stipules. **Flowers** with 5-lobed corolla and 5 stamens; **fruit** a 3-parted capsule. This small family, related to the heath family, has a single species in Alaska.

Diapensia
Diapensia lapponica L.

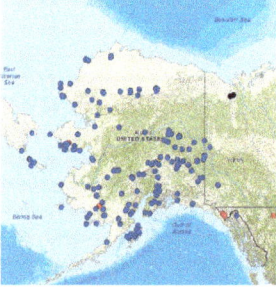

OTHER NAMES Arctic diapensia

DESCRIPTION Low, creeping, cushionlike evergreen shrub with stems horizontal or 1–3 in. (2.5–7.5 cm) high, much branched, with dense mat of dead leaves beneath.

LEAVES densely crowded and overlapping like rosettes or alternate, without stipules, narrowly oblong or spoon-shaped, 1/8–3/8 in. (3–10 mm) long, 1/16 in. (1.5 mm) wide, rounded at tip, edges turned under, thick and fleshy, stiff, hairless, dark green on upper surface, light green beneath.

TWIGS slender, hairless, concealed by leaves.

FLOWERS single, erect on stalks 1–2 in. (2.5–5 cm) high, 5/8–3/4 in. (15–20 mm) across, composed of 1–3 bracts, 5 persistent yellow green sepals, bell-shaped corolla with 5 rounded spreading lobes white or rarely pink to red, 5 alternate stamens inserted in notches of corolla, and pistil with 3-celled ovary, long slender style, and 3-lobed stigma.

FRUIT an erect elliptic or egg-shaped capsule nearly 1/4 in. (6 mm) long, 3-celled, with several seeds.

FLOWERING from late May often into July.

HABITAT Compact mats to 2 feet (60 cm) in diameter are common in dry rocky and gravelly upland slopes in arctic and alpine tundra; nearly throughout interior Alaska, south to Kodiak Island and Alaska Peninsula. Also Amchitka Island in southwestern Aleutian Islands. In southeast Alaska only in mountains above Haines and Skagway.

USES The many large flowers make diapensia showy and suitable for alpine rock gardens in spite of its small size.

DIAPENSIA - FLOWERS

DIAPENSIA

DIAPENSIA

■ ELAEAGNUS FAMILY *Elaeagnaceae*

Deciduous shrubs (elsewhere also small trees and also evergreen), covered with scurfy or star-shaped silvery or brown scales. **Leaves** alternate or opposite, simple, without stipules, not toothed on edges. **Flowers** small, lateral, single or few with both stamens and pistil (bisexual) or male and female, without petals, composed of tube (hypanthium), 4-lobed calyx, 4 or 8 stamens, and pistil with 1-celled ovary, 1 ovule, and style. **Fruit** drupelike, consisting of fleshy tube and 1–seeded nutlet.

KEY TO ALASKA ELAEAGNACEAE

1 Leaves opposite, green above and brownish scaly beneath; flowers male and female on different plants   BUFFALOBERRY (*Shepherdia canadensis*)
1 Leaves alternate, silvery scaly on both surfaces; flowers bisexual
                                                    SILVERBERRY (*Elaeagnus commutata*)

Silverberry

Elaeagnus commutata Bernh.

SYNONYM *Elaeagnus argentea* Pursh, not Calla.
DESCRIPTION Deciduous shrub 3–12 feet (1–3.7 m) high, sometimes treelike, spreading from rootstocks, much branched. LEAVES alternate, with short petiole less than 1/4 in. (6 mm) long, without stipules. **Blades** elliptic to ovate, 1–2½ in. (2.5–6 cm) long, 3/8–1 in. (1–2.5 cm) wide, blunt or short-pointed at both ends, not toothed on edges, densely silvery scaly on both surfaces, paler beneath.

TWIGS covered with rusty brown scales when young becoming silvery. Buds 1/8–1/4 in. (3–6 cm) long, covered by 2 long scales or small leaves.

FLOWERS 1–3 at base of leaves, short-stalked and turned down, 1/2–5/8 in. (12–15 mm) long, exceptionally fragrant, silvery, funnel-shaped, composed of tube with calyx of 4 yellow lobes at tip, 4 short stamens alternate and inserted in tube which is yellowish within, and pistil with 1-celled ovary, 1 ovule, and long style.

FRUIT elliptic, 1/2–5/8 in. (12–15 mm) long, silvery, drupelike, composed of dry mealy, edible tube and 1 narrow slightly 8–angled nutlet.

FLOWERING mid-June, fruit ripening in August.

HABITAT Common locally on rocky south-facing slopes and forming thickets on sandbars of major rivers in the interior.

USES The fruits are eaten raw or cooked in moosefat, especially by Alaska Natives. Plants are grown as ornamentals in interior Alaska, spreading from roots.

SILVERBERRY

SILVERBERRY

Buffaloberry
Shepherdia canadensis (L.) Nutt.

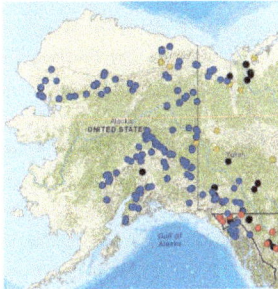

OTHER NAMES soapberry, soopolallie

SYNONYMS *Elaeagnus canadensis* (L.) A. Nels., *Lepargyraea canadensis* (L.) Greene

DESCRIPTION Deciduous shrub 2–6 feet (0.6–2 m) high, with silvery or reddish brown minute scales.

LEAVES opposite, wth short scaly petioles less than 1/8 in. (3 mm) long, without stipules. **Blades** ovate, 1/2–2 in. (1.2–5 cm) long, 1/4–1 in. (0.6–2.5 cm) wide, rounded or blunt at both ends, not toothed on edges, above green and slightly hairy with scattered star-shaped hairs, beneath densely covered with reddish brown scales and silvery, star-shaped hairs.

TWIGS gray, scaly, with paired branches, young twigs and buds covered with reddish brown scales. Buds flattened, composed of pair of small leaves (scales).

FLOWERS small, about 3/16 in. (5 mm) wide, yellowish or brownish, male and female on different plants (dioecious), in short lateral spikes in spring before the leaves, from round buds 1/16 in. (1.5 mm) in diameter formed in previous summer. **Male flowers** with calyx of 4 spreading scaly lobes and 8 stamens alternate with lobes of disk. **Female flowers** with scaly cup bearing at tip an 8-lobed disk with a 4-lobed calyx, and a pistil with 1-celled ovary and a short style.

FRUIT elliptic, red or yellowish, 1/4 in. (6 mm) long, nearly transparent, drupelike with calyx at tip, fleshy and edible but almost tasteless or bitter, and 1 nutlet. One of the earliest flowering plants

BUFFALOBERRY

in the interior, blooming in early May as soon as the snow has melted; fruits maturing in July.

HABITAT Uncommon or locally common in openings and forests of dry uplands and in aspen forests on old burns. Forming dense thickets on gravel bars of rivers near tree-line.

USES The fruits were gathered in quantities and eaten by Alaska Natives. Fruits were pressed into cakes, which were smoked and eaten, the taste sweet at first then replaced by a bitter taste (saponin) like quinine. Also the fruits were mixed with sugar and water and beat into an edible foam or froth, which was used on deserts like whipped cream. The berries are eaten in the fall by grouse and bears. Plants are sometimes grown for ornament.

BUFFALOBERRY

BUFFALOBERRY

■ HEATH FAMILY *Ericaceae*

A large family of shrubs and evergreen herbs in Alaska, usually growing in wet, acid soil, and now also encompassing former families Empetraceae (crowberry family) and Pyrolaceae (pyrola family). **Leaves** mostly alternate (sometimes basal in herbaceous species), simple, leathery, evergreen or occasionally deciduous. **Flowers** usually with funnel-shaped or urn-shaped corolla with 4 or 5 lobes, occasionally with 5 spreading distinct petals; sepals 4 or 5, partly united at base; stamens equal to or twice as many as petals; and pistil with ovary usually 5-celled, superior (except in *Vaccinium*) and 1 style. **Fruit** a capsule, berry, or drupe.

The family includes the blueberries, huckleberries, cranberries, and such showy shrubs as Labrador-tea, rhododendrons, and mountain-heaths. It also has a number of less conspicuous forest and bog shrubs, such as rusty menziesia, leatherleaf, and bog rosemary. Some examples, such as mountain-heaths, Labrador-teas, and mountain-cranberries, are evergreen, but many are deciduous. The foliage of several species is reported to be poisonous to grazing animals.

Bog-Rosemary
Andromeda polifolia L.

DESCRIPTION Small delicate, spreading, evergreen shrub, usually 1–2 feet (30–60 cm) tall, occasionally to 3 feet (1 m) often prostrate and rooting along nodes.

LEAVES narrowly elliptic to nearly linear, 1/2–1 in. (12–25 mm) long, 1/8–1/4 in. (3–6 mm) wide, thick, strongly in rolled along edges, with small projection at tip, hairless; upper surface dark green with sunken veins, lower surface whitish (glaucous); petioles short.

FLOWERS 1–4 at ends of twigs, nodding on thin reddish-purple stalks 1/4–1/2 in. (6–12 mm) long; sepals 5, short, bluntly triangular, reddish-purple; corolla pink, broadly urn-shaped, with 5 minute lobes; stamens 10.

FRUIT a spherical, 5-parted capsule, 1/8–1/4 in. (3–6 mm) in diameter, becoming erect, often persisting into winter.

FLOWERING in June and early July, fruits maturing July and August.

HABITAT Bog-rosemary is an early flowering shrub, common in bogs of the coastal and boreal forests of Alaska and in the wet sedge tundra of the northern and western parts of the state.

USES The plants contain a strong poison, andromedotoxin (from the generic name), which causes vomiting, dizziness, low blood pressure, breathing difficulty, diarrhea, and cramps. However, it is unlikely that the bitter leaves would be eaten by humans or browsed in quantity by wildlife.

BOG-ROSEMARY

BOG-ROSEMARY

BEARBERRY *Arctostaphylos*

In Alaska, low prostrate trailing or matted shrubs, although in California becoming tall shrubs and an important element in the chaparral vegetation. **Leaves** alternate, evergreen or deciduous, usually thick and leathery. **Flowers** in few-flowered racemes at the tip of stems, sepals 4 or 5 nearly separate, corolla white to pink, urn-shaped, with 4 or 5 recurved lobes, stamens 10 (sometimes 8), ovary superior, mostly 4–5-celled. **Fruit** a mealy or juicy "berry."

KEY TO ALASKA BEARBERRIES (*ARCTOSTAPHYLOS* AND *ARCTOUS*)

1   Leaves evergreen, margin entire; twigs hairy; berries red and mealy
    BEARBERRY (*Arctostaphylos uva-ursi*)

1   Leaves turning red in fall, margin toothed; twigs hairless; berries red or blue-black, juicy
    2

2   Leaves or partly skeletonized leaves persistent several years, leathery; berries blue-black
    ALPINE BEARBERRY (*Arctous alpina*)

2   Leaves dropping the first winter, thinner; berries red
    RED-FRUIT BEARBERRY (*Arctous rubra*)

Bearberry

Arctostaphylos uva-ursi (L.) Spreng.

OTHER NAMES kinnikinnik, mealberry

DESCRIPTION Prostrate evergreen shrub 3–4 in. (7.5–10 cm) tall, forming mats by rooting along the stems.

LEAVES obovate, 1/2–3/4 in. (15–20 mm) long, persistent, leathery, light green, prominently net-veined, tapering to petiole 1/8 in. (3 mm) long.

TWIGS slender and creeping, brown; bark shredding.

FLOWERS 1 to several in a raceme at ends of twigs, nodding; corolla urn-shaped, 1/8–1/4 in. (3–6 mm) long, white to pink; stalks short, 1/16–1/8 inch (1.5–3 mm).

FRUIT a red berry 1/4–5/8 in. (6–15 mm) in diameter, dry and seedy, persistent in winter.

FLOWERING in May and June, fruits ripening in August.

HABITAT Bearberry is a common shrub of dry sites in the boreal forest, usually under aspen but sometimes in open spruce stands or on open dry rocky bluffs. It often forms pure mats several yards (meters) in diameter. Common in the boreal forest region of Alaska and occasional on the north slope of the Brooks Range and Aleutian and Kodiak Islands; in southeast Alaska, in the vicinity of Glacier Bay and Lynn Canal.

USES According to reports, the dry leaves were occasionally used as a substitute for tobacco. The mealy and dry berries are rather tasteless when raw but palatable when cooked. As the name indicates, the berries are commonly eaten by bears.

BEARBERRY

BEARBERRY

BEARBERRY *Arctous*

Alpine Bearberry
Arctous alpina (L.) Nied.

OTHER NAMES ptarmiganberry, alpine ptarmiganberry
SYNONYM *Arctostaphylos alpina* (L.) Spreng.
DESCRIPTION Matted or trailing shrub 2½ –4 in. (6–10 cm) tall.
LEAVES obovate or oblanceolate, 5/8–1½ in. (15–40 mm) long and 3/8–3/4 in. (10–20 mm) wide tapering to short petiole, leathery, prominently netveined on both sides, upper side dark green, underside light green, whitish (glaucous), margins with fine teeth, skeletonized leaves remaining several years.
TWIGS prostrate, brown, hairless, with shredding bark.

FLOWERS few clustered at tips of branches, nodding; corolla 1/4–5/16 in. (6–8 mm) long, yellowish green, white, or tinged with pink.
FRUIT a juicy berry, 3/8–1/2 in. (10–12 mm) in diameter, black when ripe.
FLOWERING mid-May and June before the leaves develop, fruit ripening in August.
HABITAT Alpine bearberry is a common matted shrub of dry, wind exposed sites of the arctic and alpine tundra, and the

ALPINE BEARBERRY

treeless regions of Kodiak Island and the Aleutians. It also occurs in open black spruce stands, and dry sites in bogs at lower elevations.

USES In the fall the leaves turn a deep red and add conspicuously to the color of the tundra landscape. The berries are edible but seedy and of a rather poor taste. In poor berry years, they are often picked and mixed with blueberries. Large quantities are eaten by both bears and ptarmigan.

ALPINE BEARBERRY

ALPINE BEARBERRY

Red-Fruit Bearberry
Arctous rubra (Rehder & Wilson) Nakai

OTHER NAME ptarmiganberry

SYNONYMS *Arctostaphylos alpina* subsp. *ruber* (Rehd. & Wilson) Hultén, *Arctostaphylos rubra* (Rehd. & Wilson) Fern.

DESCRIPTION Similar to **alpine bearberry** in general appearance but somewhat taller, to 6 in. (15 cm) with red fruits.

LEAVES thinner and not as deeply wrinkled as in alpine bearberry, dropping the first winter.

FRUIT bright red when ripe, edible but seedy, with insipid taste.

HABITAT Most commonly found in lower elevation spruce forests and bogs; throughout northern Alaska and in southeastern Alaska, not in Aleutian Islands.

RED-FRUIT BEARBERRY

RED-FRUIT BEARBERRY

CASSIOPE *Cassiope*

The cassiopes are a group of white-flowered, low, prostrate, mosslike evergreen shrubs of the alpine and arctic tundra. **Leaves** scalelike or needlelike, closely pressed to stem (spreading in starry cassiope). **Flowers** with pink to white bell-shaped corolla with usually 5 (sometimes 4) short lobes; sepals usually 5 (sometimes 4), nearly separate, persistent; stamens usually 10, short.

Four species in Alaska. **Four-angled cassiope** is primarily of northern and central Alaska, the other three are restricted to mountains of Alaska Range and southward.

The members of genus *Cassiope* are often called mountain-heathers, but to distinguish them from the mountain-heaths or mountain-heathers of the genus *Phyllodoce*, it is preferable to refer to them as cassiopes.

KEY TO ALASKA CASSIOPE

1 Leaves alternate, spreading; flower 1 on short stout stalk at end of stem
STARRY CASSIOPE (*Cassiope stelleriana*)

1 Leaves opposite, pressed to stem, flowers usually 2 or more on long stalks from sides of stem   **2**

2 Leaves deeply grooved on back   **FOUR-ANGLED CASSIOPE** (*Cassiope tetragona*)

2 Leaves not grooved on back   **3**

3 Leafy stems 1/8 in. (3 mm) or more in diameter
MERTENS' CASSIOPE (*Cassiope mertensiana*)

3 Leafy stems about 1/16 in. (1.5–2 mm) in diameter
ALASKA CASSIOPE (*Cassiope lycopodioides*)

Alaska Cassiope

Cassiope lycopodioides (Pallas) D. Don

OTHER NAME clubmoss mountain-heather

DESCRIPTION Delicate low creeping mosslike evergreen shrub with erect branches only 1–2 in. (2.5–5 cm) tall.

LEAVES tiny, pressed to stem, scale-like, 1/16–1/8 in. (1.5–3 mm) long, edges with short fine hairs.

STEMS completely obscured by leaves, about 1/16 in. (1.5–2 mm) in diameter, including leaves.

FLOWERS nodding on long slender stalks 1/2–3/4 in. (12–20 mm) back from tip of stem, about 1/4 in. (6 mm) long; corolla bell-shaped, white, with usually 5 lobes nearly as long as tube; sepals rounded, reddish, transparent at edges.

FRUIT an erect round capsule about 1/8 in. (3 mm) long.

FLOWERING in June and July, fruit ripening in August.

HABITAT Arctic and alpine tundra; to 5,400 feet altitude in Juneau Ice Field.

USES The smallest of the four cassiopes in Alaska, more creeping than the others and not forming extensive mats.

ALASKA CASSIOPE

ALASKA CASSIOPE

Mertens' Cassiope
Cassiope mertensiana (Bong.) D. Don

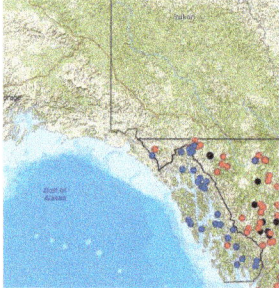

OTHER NAMES Mertens' mountain-heather, white heather
DESCRIPTION Prostrate, mosslike, evergreen, mat-forming shrub with upturned branches 2–12 in. (5–30 cm) tall.
LEAVES opposite in 4 rows and pressed to stem, scalelike, ovate-lanceolate, 1/16–1/8 in. (1.5–3 mm) long, rounded on back and grooved at base, hairless or with small glandular hairs along edge.
STEMS covered by leaves except at base, including leaves about 1/16–1/8 in. (1.5–3 mm) in diameter, 4-angled or square in cross section.
FLOWERS several about 1/4 in. (6 mm) long on slender erect stalks 1/4–1 in. (6–25 mm) long, from sides of stems; corolla bell-shaped, white to pink, with 5 lobes about half as long as tube; sepals 5, rounded, reddish.
FRUIT an erect round capsule 1/8 in. (3 mm) in diameter.
FLOWERING from late June through August, fruits maturing August and September.
HABITAT A common mat-forming shrub in seepage areas, protected slopes, mountain meadows, and slopes adjacent to snowfields in southeastern Alaska to about 5,000 feet altitude, where it is usually associated with related species of *Cassiope*.
ETYMOLOGY This species honors Carl Heinrich Mertens (1796–1830), German naturalist, who discovered it at Sitka in 1827.

MERTENS' CASSIOPE

MERTENS' CASSIOPE

Starry Cassiope
Cassiope stelleriana (Pall.) DC.

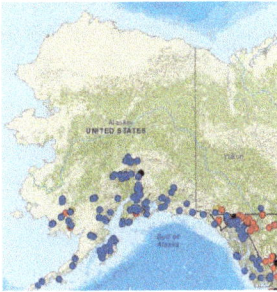

OTHER NAMES moss heather, Alaska heather, Alaska moss heath

SYNONYM *Harrimanella stelleriana* (Pall.) Cov.

DESCRIPTION Low spreading, mat-forming evergreen shrub, the upright stems from 2–4 in. (5–10 cm) tall.

LEAVES alternate, spreading, linear-lanceolate, 1/8–3/16 in. (3–5 mm) long, keeled on lower surface, hairless or with hairs along edge.

STEMS slender, reddish, with scattered short stiff hairs.

FLOWERS usually 1, about 1/4 in. (6 mm) long, erect to nodding at end of stem on stout stalk 1/8–3/8 in. (3–10 mm) long; corolla bell-shaped, white to pink; sepals rounded, reddish, united in lower half.

FRUIT an erect round capsule about 1/8 in. (3 mm) long.

FLOWERING late May through July, fruits maturing late July and August.

HABITAT A trailing mat-forming shrub of protected slopes, snow deposition areas, and moist seepage areas in alpine areas of southeastern and south central Alaska. It is com-

STARRY CASSIOPE

mon in southeastern Alaskan mountains, usually associated with mountain-heaths; to 5,400 feet (1,650 m) on rocky cliffs of Juneau Ice Field. It is uncommon in the Alaska Range, where often growing with four-angled cassiope.

ETYMOLOGY The scientific name commemorates Georg Wilhelm Steller (1708–46), German naturalist with Bering's expedition, who in 1741 made the first plant collections in Alaska.

STARRY CASSIOPE

Four-Angled Cassiope
Cassiope tetragona (L.) D. Don

OTHER NAMES firemoss cassiope, four-angled mountain-heather, Lapland cassiope

DESCRIPTION Low, creeping, evergreen, mosslike mat-forming shrub with stems 4–8 in. (1–2 dm) tall.

LEAVES opposite in 4 rows, thick, lance-shaped, 1/8–3/16 in. (3–5 mm) long, deeply grooved, with short fine hairs along edge.

STEMS covered by leaves except at base, 4-angled, including leaves nearly 3/16 in. (5 mm) in diameter.

FLOWERS about 1/4 in. (6 mm) long, nodding on slender stalks 1/2–1 in. (12–25 mm) from sides of upper 1½ in. (4 cm) of stem; corolla bell-shaped, white, with lobes about half as long as tube; sepals separate nearly to base, rounded and reddish.

FRUIT an erect round capsule 1/8 in. (3 mm) long.

FLOWERING from late May through August, fruit maturing August and September.

HABITAT One of the prettiest and most common of the cassiopes on the alpine and arctic tundra of central and northern Alaska. It forms rather dense mats in protected areas that are snow covered during the winter and that have snow remaining into the summer.

NOTE sometimes called firemoss because even in the green condition it burns rather well and has been used by mountain climbers and arctic travelers as a source of fuel in areas where no larger woody plants are available.

FOUR-ANGLED CASSIOPE

FOUR-ANGLED CASSIOPE

Leatherleaf
Chamaedaphne calyculata (L.) Moench

OTHER NAME cassandra

SYNONYM *Cassandra calyculata* (L.) D. Don.

DESCRIPTION Prostrate to erect evergreen shrub, rooting at nodes, usually 2–3 feet (6–10 dm) tall.

LEAVES alternate, oblong to elliptic, 1/2–1¼ in. (12–30 mm) long and 1/4–1/2 in. (6–12 mm) wide, thick, leathery, and slightly rolled downward on edges, surfaces dark green with scurfy scales often appearing as white dots; petioles short.

TWIGS with fine short white hairs when young but becoming hairless with age, light to dark brown.

FLOWERS several to many in a row on short stalks, hanging down from lower side of stem (a leafy raceme) about 1/4 in. (6 mm) long, corolla white, cylindrical and slightly constricted just below the 5 short triangular lobes; sepals 5, thick, green ovate to lanceolate, with dense hairs on margins; stamens 10, short.

FRUIT a round 5-parted capsule about 1/8 in. (3 mm) in diameter, longer than sepals with slender style persistent.

HABITAT An abundant shrub in bogs and open black spruce stands throughout the boreal forest. North of the tree-line, it is rare and occurs primarily in wet sites along river terraces.

NOTE Leatherleaf is one of the earliest flowering plants in the interior of Alaska, flowering in early to late May, usually before leaves of most plants have developed. In fall, winter, and spring, the leaves have a reddish color, giving many bogs this hue when viewed from a distance.

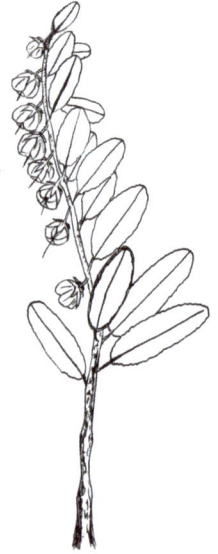

LEATHERLEAF

LEATHERLEAF

LEATHERLEAF - FLOWERS

Pipsissewa

Chimaphila umbellata (L.) W. Barton

OTHER NAMES princes-pine, wintergreen, waxflower
SYNONYM *Chimaphila occidentalis* Rydb.
DESCRIPTION A low evergreen half-shrub, clumped or matlike, with creeping branches that ascend to 4–12 in. (10–30 cm).
LEAVES thick, shiny, 3/4–2½ in. (3–7 cm) long, 3/16–1 in. (0.5–2.5 cm) wide, broadest near tip, tapering toward base into a short petiole 1/8–5/16 in. (3–8 mm) long, sharply toothed, alternate or whorled on the stem.
TWIGS slender, only semi-woody, yellow or green.
FLOWERS 4 to 15, nodding in a cluster at the end of the twigs, on a stalk 2–4 in. (5–10 cm) long, saucer-shaped; petals separate, reddish to pink, 3/16–5/16 in. (5–8 cm) long, sepals hairy, fringed at tip.
FRUIT a spherical dry, 5-parted, many seeded capsule 1/4–5/16 in. (6–8 mm) in diameter.
HABITAT an uncommon shrub in southeastern Alaska, growing under Sitka spruce and hemlock.

PIPSISSEWA

PIPSISSEWA

Copperbush
Elliottia pyroliflora (Bong.) Brim & P.F. Stevens

OTHER NAME copper-flower
SYNONYM *Cladothamnus pyroliflorus* Bong.
DESCRIPTION Erect shrub 1½–4½ feet (0.5–1.5 m) tall, with clustered long leaves and showy copper-colored flowers.
LEAVES ovate to oblanceolate, 3/4–1½ in. (20–40 mm) long, 3/16–1/2 in. (5–12 mm) wide, with rounded to abruptly pointed (mucronate) tip, pale green and somewhat whitish (glaucous) on underside, appearing in whorls.
FIRST YEAR TWIGS light brown and shiny, stiff, 1/16 inch (1.5 mm) in diameter, with minute hairs, older twigs with shredding bark. **Buds** asymmetrical, short-pointed, orange, shiny, of 2 keeled scales.
FLOWERS 1 to several at ends of twigs, about 1 in. (25 mm) across; sepals 5, narrow; 5 spreading oval copper-colored petals 3/8–5/8 in. (10–15 mm) long; stamens 10, 3/8 inch (10 mm) long, hooked near tip; style long, curved.
FRUIT a round capsule 1/8–1/4 in. (3–6 mm) in diameter, dark reddish brown. Flowering from late June through middle of August, fruits ripening August and September.
HABITAT Copperbush forms dense clumps several yards (meters) across in meadows at and just above tree-line and in openings and along streambanks within the coastal forests.
USES The unusual color of the flowers makes this shrub desirable for cultivation. It is often planted in southeastern Alaskan towns.

COPPERBUSH

COPPERBUSH

Crowberry
Empetrum nigrum L.

OTHER NAMES mossberry, blackberry, curlewberry
SYNONYM *Empetrum hermaphroditum* (Lange) Hagerup
DESCRIPTION Low, creeping or spreading evergreen heather-like shrub to 6 in. (15 cm) high, forming dense mats, with horizontal, much branched stems.
LEAVES crowded, 4 in a whorl or sometimes alternate, without stipules, with minute petiole, linear or needlelike, 1/8–1/4 in. (3–6 mm) long, shiny yellow green, with groove on lower surface formed by curved margins, hairless.
TWIGS curving upward 2–6 in. (5–15 cm) long, very slender, brown, finely hairy, becoming shreddy.
FLOWERS single, inconspicuous, stalkless at base of leaves, small, 1/8–1/4 in. (3–6 mm) long, purplish, composed of 3 bracts, 3 sepals, 3 spreading petals, 3 stamens much longer than petals, and pistil with 6–9-celled ovary and flat stigma with 6–9 narrow lobes; also some plants with male flowers and others with female flowers.
FRUIT round, berrylike, 3/16–3/8 in. (5–10 mm) or more in diameter, shiny dark blue black or purple, very juicy and sweet, containing 6–9 reddish brown nutlets.
FLOWERING in June, fruits ripening in August and persisting under the snow throughout the winter.
HABITAT Common and widespread in arctic-alpine tundra, moist rocky slopes, and muskegs, also in spruce forests, almost throughout Alaska including Aleutian Islands. One of the commonest species in heath mats to 5,600 feet altitude on the rocky cliffs or nunataks of the Juneau Ice Field. In the interior, mostly in mountains, also along southern coast.
USES The edible berries are consumed in quantities locally, usually mixed with other berries and reported to be excellent in pies. In winter Eskimos gather the fruits under the snow. The berries serve also as fall and winter food of grouse, ptarmigan, and bear. Some plants bear fruits in abundance, but male plants have none. Crowberry is planted as a ground cover in rough low areas in interior Alaska. Plants can be grown from cuttings.

CROWBERRY

CROWBERRY

CROWBERRY

WINTERGREEN *Gaultheria*

Low evergreen shrubs. **Leaves** alternate, evergreen, ovate to elliptic, toothed on edges. **Flowers** with urn-to bell-shaped pink corolla with 5 short lobes; calyx with 5 short glandular hairy lobes; stamens 10, short. **Fruit** a berrylike, fleshy, 5-celled capsule surrounded by the enlarged fleshy calyx.

KEY TO ALASKA WINTERGREENS

1 Leaves 2–4 in. (5–10 cm) long, sharply toothed; flowers many; fruit purplish; low shrub of southeast Alaska **SALAL** (*Gaultheria shallon*)
1 Leaves 5/8–1 3/8 in. (15–35 mm) long, finely wavy toothed; flowers 1–6; fruit white; prostrate shrub of Kiska Island in eastern Aleutians

MIQUEL WINTERGREEN (*Gaultheria pyroloides*)

Miquel Wintergreen

Gaultheria pyroloides Hook. f. & Thomson ex Miquel

SYNONYM *Gaultheria miqueliana* Takeda

DESCRIPTION Low, prostrate evergreen shrub to 16 in. (41 cm) high.

LEAVES oval, 5/8–1 3/8 in. (15–35 mm) long, and 3/8–5/8 in. (10–15 mm) wide, wavy toothed, rounded at tip.

FLOWERS 1–6, about 1/4 in. (6 mm) long, in glandular hairy racemes; corolla urn-shaped, pink; calyx lobes triangular, glandular hairy on back.

FRUIT a fleshy white berrylike capsule.

HABITAT Miquel wintergreen, a small Asiatic shrub, has been collected only on Kiska Island in the western Aleutians, and illustrates the close relationship between the flora of eastern Asia and western Alaska; also in eastern Asia and Japan.

ETYMOLOGY Named for Frederick Anton Willem Miquel (1811–71), Dutch botanist.

MIQUEL WINTERGREEN

Salal
Gaultheria shallon Pursh

DESCRIPTION Stiff, creeping to erect evergreen shrub, 2–3 feet (0.6–1 m) tall.

LEAVES alternate, short-stalked, large, thick, ovate to elliptic, 2–4 in. (5–10 cm) long, 1–2 in. (2.5–5 cm) wide, stiff and leathery, short-pointed at tip, sharply toothed on edges, with occasional long reddish hairs, upper surface shiny green with raised veins, lower surface lighter green.

TWIGS with scattered long, often gland-tipped hairs, hairless with age, reddish-brown, with shredding bark.

FLOWERS 5–15 in long glandular hairy racemes, usually at tips of twigs, 3/8 in. (1 cm) long; corolla urn-to bell-shaped, 5/16–3/8 in. (8–10 mm) long, pink, with stiff reddish brown hairs, and 5 short triangular lobes; calyx lobes reddish-brown, glandular haired, about 1/3 as long as corolla; stamens 10 short.

FRUIT a round capsule enclosed by fleshy calyx, berrylike, purplish, bristly hairy, 1/4–1/2 in. (6–12 mm) in diameter. Collected in flower in May and June.

HABITAT Salal is a common undershrub of poor scrub timber sites of western redcedar, Alaska-cedar, spruce and hemlock forests in southern parts of southeastern Alaska; it may form a nearly continuous cover in some stands.

USES The spicy berries are eaten by grouse and other birds but seldom by humans. It is reported that Native Americans in the Northwest gathered the fruits.

NOTE The stiff evergreen leaves and densely hairy flowers and twigs make this shrub easily recognized.

SALAL - FLOWERS

SALAL - FRUIT

MOUNTAIN-LAUREL *Kalmia*

Low or prostrate evergreen shrubs. **Leaves** opposite, leathery. **Flowers** single or several in terminal or axillary clusters. **Fruit** a capsule.

KEY TO ALASKA *KALMIA*

1 Stamens 5; matted dwarf shrub   **ALPINE-AZALEA** (*Kalmia procumbens*)
1 Stamens 10; erect, branched dwarf shrub   **BOG KALMIA** (*Kalmia polifolia*)

Bog Kalmia
Kalmia polifolia Wangenh.

OTHER NAMES swamp-laurel, bog-laurel, pale-laurel, alpine-azalea

SYNONYM *Kalmia microphylla* (Hook.) Heller

DESCRIPTION Evergreen spreading shrub of bogs and mountain meadows, 4–20 in. (1–5 dm) tall, with showy rose to purple flowers.

LEAVES opposite, stalkless, oblong to linear 3/4–1½ in. (2–4 cm) long, 1/8–5/16 in. (3–8 mm) wide, flat or with edges rolled under, dark green above and whitish (glaucous) beneath; petioles short, 1/16–3/16 in. (1.5–5 mm) long.

TWIGS slightly 2–angled.

FLOWERS several in cymes at ends of twigs on stalks 3/8–1¼ in. (1–3 cm) long; corolla saucer-shaped, 3/8–3/4 in. (1–2 cm) across, with 5 lobes and 10 ridges (keels), rose to purple; sepals 5, thick; stamens 10.

FRUIT a 5-parted capsule about 3/16 in. (5 mm) long.

FLOWERING from late May to early July, fruits maturing in August.

HABITAT Occasional in wet open habitats of mountains and lowlands in southeast Alaska.

BOG KALMIA

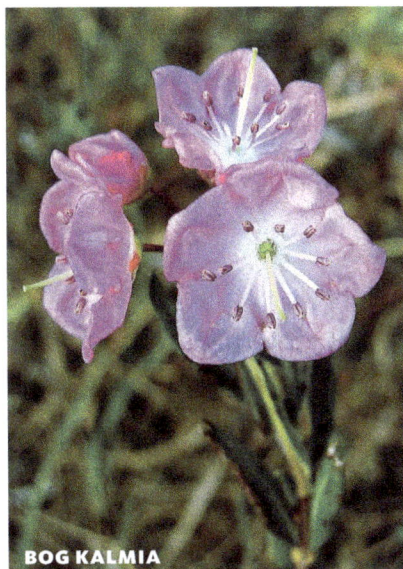

BOG KALMIA

Alpine-Azalea

Kalmia procumbens (L.) Gift, Kron & P.F. Stevens ex Galasso, Banfi & F. Conti

SYNONYM *Loiseleuria procumbens* (L.) Desv.

DESCRIPTION Matted or trailing evergreen subshrub 1–2 in. (25–50 mm) tall.

LEAVES opposite, elliptic, small, 1/8–1/4 in. (3–6 mm) long and 1/16 in. (1.5 mm) wide, leathery, with margins rolled under, upper side hairless, lower side with dense short white hairs and a prominent ridge.

TWIGS much branched, nearly totally concealed by the persistent leaves.

FLOWERS 1 to several at ends of twigs, erect on stalks 1/8–1/4 in. (3–6 mm) long; corolla bell-shaped, pink or sometimes white, 1/8–3/16 in. (3–5 mm) long, divided nearly to the middle into 5 lobes; calyx deeply divided into 5 reddish-purple lanceolate lobes; stamens 5.

FRUIT an erect, round 2–3-parted dark red capsule 1/8–3/16 in. (3–5 mm) in diameter.

FLOWERING from late May through July, fruits maturing in July and August.

HABITAT occasional to common on well drained rocky sites in arctic and alpine tundra throughout most of Alaska. It frequently forms pure mats, usually 4–12 in. (1–3 dm) in diameter, but as wide as 1 yard (1 m) with a large number of flowers relative to the size of the plant.

ALPINE-AZALEA

ALPINE-AZALEA

ALPINE-AZALEA - FRUITING

Rusty Menziesia
Menziesia ferruginea Sm.

OTHER NAMES skunkbrush, fool's-huckleberry

SYNONYM *Rhododendron menziesii* Craven

DESCRIPTION Loose-spreading, odorous, deciduous shrub to 6–10 feet (2–3 m) high, with slender, widely forking paired branches and small yellowish red flowers.

LEAVES obovate to elliptic, $1\frac{1}{4}$–$2\frac{1}{2}$ in. (3–6 cm) long and 1/2–3/4 in. (1.2–2 cm) wide, short-pointed usually with abrupt (mucronate) tip, edges minutely toothed with gland-tipped hairs, upper side gray green with scattered brown hairs, under side whitish (glaucous) with glandular ("sticky") hairs; petioles 1/8 in. (3 mm) with gland-tipped hairs.

YOUNG TWIGS glandular, with odor when crushed, older twigs reddish brown to gray, smooth to peeling in thin layers. **Buds** of 2 sizes, the larger with many scales developing into flower cluster.

FLOWERS several to many at ends of twigs on glandular stalks 3/8–3/4 in. (1–2 cm) long; corolla urn-shaped, yellowish red (sometimes described as coppery-pink), 1/4–1/2 in. (6–12 mm) long, with 4 shallow lobes; calyx 4-lobed, with long glandular hairs; stamens 8; stigma 4-lobed.

FRUIT an ovoid, 4-parted capsule 3/16–5/16 in. (5–8 mm) long, green to reddish brown, often persistent through the winter.

FLOWERING from late May through July, capsules maturing July and August.

HABITAT A common shrub in undergrowth of the coastal spruce-hemlock forest, often under a dense canopy, also in openings and on cutover forest land, especially on well drained slopes in association with blueberries. It also grows in the southern part of the boreal forest in white spruce and white spruce-paper birch stands.

NOTE Because of the leaf and flower size and shape, menziesia is sometimes confused with the huckleberries (*Vaccinium*), but its fruit is a capsule rather than a berry.

ETYMOLOGY This genus was dedicated to Archibald Menzies (1754–1842), Scotch physician and naturalist with Vancouver's voyage of 1793–94 to the Northwest coast.

RUSTY MENZIESIA

RUSTY MENZIESIA

RUSTY MENZIESIA

MOUNTAIN-HEATH *Phyllodoce*

Low, clump or mat-forming evergreen shrubs of alpine tundra. **Leaves** crowded, small, linear, blunt-pointed. **Twigs** with conspicuous peglike leaf-scars. **Flowers** in terminal clusters (corymbs); corolla bell-shaped or urn-shaped with 5 small lobes; sepals 5, persistent; stamens 10, short. **Fruit** a 5-parted rounded capsule.

KEY TO ALASKA MOUNTAIN-HEATHS

1 Corolla bell-shaped, not constricted at mouth; flowers pink to red
                                            RED MOUNTAIN-HEATH (*Phyllodoce empetriformis*)
1 Corolla urn-shaped, constricted at mouth; flowers yellow or blue   **2**
2 Corolla purple to blue   **BLUE MOUNTAIN-HEATH** (*Phyllodoce caerulea*)
2 Corolla yellow   **3**
3 Corolla glabrous on outside   **ALEUTIAN MOUNTAIN-HEATH** (*Phyllodoce aleutica*)
3 Corolla glandular on outside; filaments more or less pubescent at base
                               YELLOW MOUNTAIN-HEATH (*Phyllodoce glanduliflora*)

Aleutian Mountain-Heath

Phyllodoce aleutica (Spreng.) Heller

OTHER NAMES Aleutian mountain-heather, Aleutian heather, cream mountain-heather, yellow mountain-heather, yellow heather

DESCRIPTION Low much branched, yellow-flowered evergreen shrub, 2–6 in. (5–15 cm) tall.

LEAVES needlelike, linear, thick, 1/4–1/2 in. (6–12 mm) long, 1/16 in. (1.5 mm) wide, with minute glandular teeth on edge, yellow-green, grooved, hairy on lower surface, crowded in upper 2–4 in. (5–10 cm) of stem.

STEMS much branched, slender, with conspicuous peglike leaf-scars.

FLOWERS 5–15 at tips of erect or nodding stems, glandular hairy stalks, 1/2–5/8 in. (12–15 mm) long; corolla yellow-green, urn-shaped, 1/4 in. (6 mm) long, with 5 small lobes, hairless, calyx with short-pointed lobes divided nearly to base, hairless or glandular hairy.

FRUIT a capsule 5/16–3/8 in. (8–10 mm) long, oval, splitting into 5 parts. Commonly blooming from early June until late August.

HABITAT Mountains along coast of south-eastern Alaska westward to western Aleutians, and along the west coast as far north as Yukon River, both above and below timberline. It forms pure

**ALEUTIAN
MOUNTAIN-HEATH**

mats several yards (meters) in diameter, especially at the head of snow field slopes. In the mountains near Juneau, it forms extensive heath mats to elevations of 5,400 feet; also at sea level on fresh moraine and outwash deposits.

ALEUTIAN MOUNTAIN-HEATH

ALEUTIAN MOUNTAIN-HEATH

Blue Mountain-Heath
Phyllodoce caerulea (L.) Bab.

OTHER NAME blue mountain-heather
DESCRIPTION Low matted evergreen shrub 2–6 in. (5–15 cm) high, with purple or blue flowers.
LEAVES scattered, needlelike, linear, 1/8–1/4 in. (3–6 mm) long, 1/16–3/16 in. (1.5–5 mm) wide, rounded at tip, shiny dark green, hairless, grooved on undersurface.
STEMS much branched, slender, with conspicuous peglike leaf-scars and shredded bark.
FLOWERS 3–4 at tips of stems on erect to curved glandular stalks 3/8–5/8 in. (1–1.5 cm) long; corolla urn-shaped with 5 small lobes, 5/16–3/8 in. (8–10 mm) long, purple to blue.
FRUIT an oval capsule, 1/16–1/8 in (1.5–3 mm) long, erect on stalk elongating in fruit to 1 in. (2.5 cm).
FLOWERING in July and August, fruits maturing in August and September.
HABITAT Uncommon shrub of the coastal and mountain tundra of central and western Alaska, usually in depressions where the snow remains late in the spring.

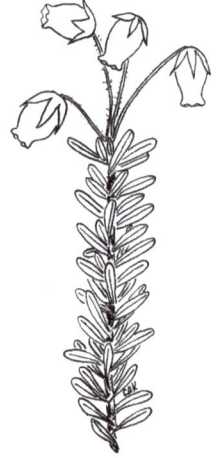

BLUE MOUNTAIN-HEATH

BLUE MOUNTAIN-HEATH

BLUE MOUNTAIN-HEATH

Red Mountain-Heath
Phyllodoce empetriformis (Sm.) D. Don

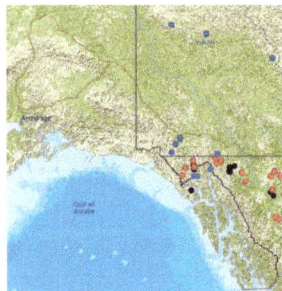

OTHER NAMES red mountain-heather, red heather, purple heather

DESCRIPTION Low matted evergreen shrub 4–8 in. (1–2 dm) tall, with pink to red flowers. **LEAVES** linear, 1/4–1/2 in. (6–12 mm) long and 1/16 in. (1.5 mm) wide, crowded on the upper 2–4 inches (5–10 cm) of stem, edges with minute glandular teeth, with 2 deep grooves on lower surface.

STEMS slender, gray, with conspicuous peglike leaf-scars.

FLOWERS 5–15 at tips of stems on slightly nodding to upright glandular-haired stalks 5/8–3/4 in. (15–20 mm) long; corolla pink to red, bell-shaped, 3/16–1/4 in. (5–6 mm) long, divided 1/4 into 5 lobes which are rolled outwards; sepals 5, divided nearly to base, dark red, persistent.

FRUIT an erect capsule, 3/16–1/4 in. (5–6 mm) long. Collected in flower in early and late August.

HABITAT Uncommon alpine or subalpine shrub in southeast Alaska in protected snow deposition areas where it usually occurs with one or more of the other mountain-heaths.

RED MOUNTAIN-HEATH

RED MOUNTAIN-HEATH

Yellow Mountain-Heath
Phyllodoce glanduliflora (Hook.) Coville

SYNONYM *Phyllodoce aleutica* subsp. *glanduliflora*

DESCRIPTION Low-growing matted shrub with much-branched, erect stems 4–8 in. (10–20 cm)

LEAVES alternate, spreading-ascending, 1/8–3/8 in. (3–9 mm) long, linear, dark green, with edges rolled under, the margin minutely glandular-serrate.

TWIGS hairy.

FLOWERS solitary, nodding, on hairy pedicels mostly longer than 1 cm when in flower, greenish yellow; corolla urn-shaped, bright yellow, 1/3 in. (9 mm) long, stipitate-glandular on outer surface, with 5 small spreading lobes; flowers and stalks sticky with glands; sepals 1/8–1/4 in. (3–6 mm) long, greenish, stipitate-glandular.

HABITAT Subalpine and alpine tundra on rocky sites. Also found in lower washes and at sea level in Prince William Sound.

NOTE This attractive plant is known to hybridize with *P. empetriformis*. The hybrids are known as *P.* x *intermedia* (Hook.) Camp. *P. granduliflora* also apparently forms hybrids with *P. aleutica* in south-central Alaska.

TIP Very similar to *Phyllodoce aleutica*, but corolla glandular on outside. Leaves similar to those of *P. caerulea* but shorter, yellowish green, the obscure marginal serrulations glandular hairy.

YELLOW MOUNTAIN-HEATH

YELLOW MOUNTAIN-HEATH

RHODODENDRON *Rhododendron*

Low evergreen shrubs or sub-shrubs in Alaska (elsewhere also tall shrubs and small trees). **Leaves** alternate, entire on margins, with short petioles. **Flowers** with showy corolla with 5 large spreading lobes, calyx 5-parted and small, stamens 10, and long slender persistent style. **Fruit** an oblong capsule mostly 5-parted.

KEY TO ALASKA RHODODENDRON

1 Leaves not rusty-tomentose below   **LAPLAND ROSEBAY** (*Rhododendron lapponicum*)
1 Leaves densely rusty-tomentose below   **2**
2 Leaves nearly linear, tightly rolled under, 1/32–1/16 in. (0.8–1.5 mm) wide; stalks of flower and fruit abruptly bent just below tip
                    NARROW-LEAF LABRADOR-TEA (*Rhododendron tomentosum*)
2 Leaves wider, 3/16–1/2 in. (5–12 mm), edges slightly rolled under, flower stalks evenly curved   **LABRADOR-TEA** (*Rhododendron groenlandicum*)

Labrador-Tea

Rhododendron groenlandicum (Oeder) Kron & Judd

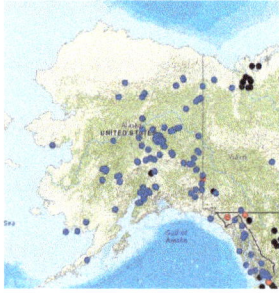

SYNONYMS *Ledum groenlandicum* Oeder, *Ledum palustre* L. subsp. *groenlandicum* (Oeder) Hultén

DESCRIPTION Evergreen shrub 3 feet (1 m) tall, with upright or prostrate branches.

LEAVES narrowly oblong, 1–2 in. (25–50 mm) long, 3/16–1/2 in. (5–12 mm) wide, with fragrant odor, leathery, margins strongly rolled under, underside covered with reddish brown woolly hairs, upper surface dark green and roughened.

YOUNG TWIGS hairy, light brown, older twigs gray.

FLOWERS numerous, conspicuous, white, fragrant, in clusters at end of twigs, 5/8 in. (15 mm) across; petals 5, spreading 3/16–5/16 in. (5–8 mm) long; stamens mostly 8; flower stalks 3/8–3/4 in. (10–20 mm) long, evenly curved.

FRUIT a hairy, oblong capsule 1/4 in. (6 mm) long, opening from base in autumn and persistent most of winter.

HABITAT Common shrub of black spruce and birch forests and bogs. It colonizes abundantly after fire in the black spruce type. It is also abundant near tree-line in open white spruce stands, where it blooms profusely from mid-June to mid-July. In southeast Alaska it grows in open bogs at low elevations.

USES As with narrow-leaf Labrador-tea, a tea can be made by boiling the aromatic dried leaves, though seldom used today.

LABRADOR-TEA

LABRADOR-TEA

Lapland Rosebay
Rhododendron lapponicum (L.) Wahlenb.

OTHER NAME alpine rhododendron

DESCRIPTION Matted to erect, much branched evergreen shrub, 4–16 in. (1–4 dm) tall, with showy purple flowers.

LEAVES oval, 3/16–9/16 in. (5–15 mm) long, 1/8–1/4 in. (3–6 mm) wide, blunt at tip, somewhat rolled down on margins, leathery, crowded at ends of twigs, both surfaces with greenish to brown resin dots, new leaves light green, old leaves dark green to brown.

TWIGS stout, much branched, scurfy with resin dots.

FLOWERS 1 to several in terminal clusters, fragrant, corolla spreading and slightly irregular, pinkish to deep purple, 5/8–3/4 in. (15–20 mm) across; stalks 1/4–1/2 in. (6–12 mm) long, scurfy, curved or straight.

FRUIT a dry capsule 3/16–1/4 in. (5–6 mm) long, opening from tip, persisting through most of winter.

HABITAT Occasional to rare, early-flowering shrub of tundra and open spruce forests at tree-line.

NOTE When flowering in late May to mid-June, the fragrance of this plant becomes noticeable. Individual shrubs and flowers are showy, but the shrub is rarely abundant.

LAPLAND ROSEBAY

LAPLAND ROSEBAY

Narrow-Leaf Labrador-Tea

Rhododendron tomentosum Harmaja

OTHER NAMES Hudson Bay tea

SYNONYMS *Ledum decumbens* (Ait.) Lodd., *Ledum palustre* L. subsp. *decumbens* (Ait.) Hultén

DESCRIPTION Evergreen shrub 1–2 feet (3–6 dm) tall, similar to the more common Labrador-tea but plants smaller and with narrower leaves rolled under at edges.

LEAVES linear, 5/16–5/8 in. (8–15 mm) long, 1/16–1/8 in. (1.5–3 mm) wide, leathery, rolled under at edges, upper surface shiny, dark green, lower surface with reddish-brown woolly hairs.

YOUNG TWIGS hairy, light brown, older twigs gray.

FLOWERS numerous, in clusters at tips of twigs, about 1/2 in. (12 mm) broad; petals 5, white, spreading 3/16–5/16 in. (5–8 mm) long; stamens mostly 10; flower stalks 1/2–3/4 in. (12–20 mm) long, sharply and abruptly bent just below tip.

FRUIT a capsule 1/8–1/4 in. (3–6 mm) long, oval, finely hairy, maturing in July and August, opening from base in autumn and persistent most of winter.

HABITAT A common shrub in arctic and alpine tundra in sedge tussocks and wet depressions. In the boreal forest it is common in sphagnum bogs and wet black spruce types.

USES A palatable tea can be made by boiling the aromatic leaves of either species of Labrador-tea. However, if used in large quantities, it may have a cathartic effect.

NOTE The large white, fragrant flower clusters of this shrub are conspicuous during June and early July.

NARROW-LEAF
LABRADOR-TEA

NARROW-LEAF LABRADOR-TEA

KAMCHATKA RHODODENDRON *Therorhodion*

Small, evergreen subshrubs (ours). **Leaves** alternate; petiole absent or nearly so. **Flowers** in terminal racemes, each flower in a bract axil, bilaterally symmetric; sepals 5; petals 5; stamens 10; ovary 5-chambered; style long and curved. **Fruit** a capsule. Our species formerly included within the closely related genus *Rhododendron.*

KEY TO ALASKA *THERORHODION*

1 Leaves distinctly stipitate-glandular; corolla ± glabrous externally; plants of the Seward Peninsula and lower Yukon River Valley

GLANDULAR KAMCHATKA RHODODENDRON (*Therorhodion glandulosum*)

1 Leaves lacking stipitate-glandular hairs (or sparsely stipitate-glandular); corolla pubescent externally; plants of coast and islands of southern Alaska

KAMCHATKA RHODODENDRON (*Therorhodion camtschaticum*)

Kamchatka Rhododendron
Therorhodion camtschaticum (Pall.) Small

SYNONYM *Rhododendron camtschaticum* Pall.

DESCRIPTION Low, evergreen subshrub 2–6 in. (5–15 cm) tall with large showy flowers.

LEAVES obovate, 1/2–1¾ in. (12–45 mm) long and 3/8–3/4 in. (10–20 mm) wide, tapering to base, with conspicuous stiff hairs on margins and prominent network of veins on underside; petiole lacking.

TWIGS coarse, much-branched, gray brown to reddish, with shredding bark.

FLOWERS 1 to several on erect leafy stalks 3/4–1¼ in. (2–3 cm) long at ends of twigs; corolla rosepurple to deep red, spreading, 1¼–1¾ in. (32–45 mm) across, style conspicuous, 1/2–3/4 in. (12–20 mm) long, curved.

FRUIT a capsule 1/4–3/8 in. (6–10 mm) long on a long stalk.

HABITAT Common in some areas of the Aleutian Islands, especially on dry rocky tundra characterized by the heath family. In forested regions, it is a low shrub of the alpine zone.

ADDITIONAL SPECIES Glandular Kamchatka rhododendron (*Therorhodion glandulosum*

Standl. ex Small, synonym *Rhododendron camtschaticum* ssp. *glandulosum* (Standl. ex Small) Hultén) is similar to *T. camtschaticum* but the corolla is without hairs on its outside and the leaf margins have glandular

KAMCHATKA RHODODENDRON

hairs. In *T. camtschaticum*, the corolla is hairy on the outside and the leaf margins mostly with non-glandular hairs.

KAMCHATKA RHODODENDRON

BLUEBERRY *Vaccinium*

OTHER NAMES huckleberry, mountain-cranberry, cranberry

Low creeping, or tall ascending shrubs, mostly deciduous but sometimes evergreen. **Leaves** alternate, often leathery. **Flowers** 1 to several at base of leaves or at ends of twigs; corolla urn-shaped or bell-shaped with 4–5 lobes or of 4 petals bent backward; calyx of 4–5 persistent teeth or lobes on inferior ovary; stamens 8–10, within corolla. **Fruit** a blue or red, round juicy berry.

All Alaskan *Vaccinium* produce edible fruit. Only three, bog blueberry, mountain-cranberry, and bog cranberry, reach northern Alaska, the rest are primarily species of the coastal forest.

KEY TO ALASKA *VACCINIUM*

1   Leaves evergreen, thick; low trailing shrubs   2
1   Leaves deciduous, corolla urn-shaped; usually upright shrubs, though occasionally rooting at nodes   3
2   Leaves oval; corolla bell-shaped   MOUNTAIN-CRANBERRY (*Vaccinium vitis-idaea*)
2   Leaves lance-shaped; corolla of 4 petals bent backward
                                                       BOG CRANBERRY (*Vaccinium oxycoccos*)
3   Twigs round; plants usually less than 16 in. (40 cm) tall   4
3   Twigs angled; plants usually more than 2 ft (6 cm) tall   5
4   Leaves entire on margins; flowers 1–4 from scaly buds on old twigs
                                                      BOG BLUEBERRY (*Vaccinium uliginosum*)
4   Leaves finely toothed on margins; flower 1 on new twig
                                                    DWARF BLUEBERRY (*Vaccinium cespitosum*)
5   Fruit red; leaves usually less than 1 in. (2.5 cm) long; twigs green, strongly angled
                                                   RED HUCKLEBERRY (*Vaccinium parvifolium*)
5   Fruit blue or black; leaves commonly more than 1 in. (2.5 cm) long; twigs reddish to brown, weakly angled   6
6   Flowering with or before the leaves; corolla longer than broad; stalk not enlarged below fruit; leaves without hairs on midrib beneath   EARLY BLUEBERRY (*Vaccinium ovalifolium*)
6   Flowering after the leaves; corolla as broad or broader than long; stalk enlarged just below fruit; leaves with fine hairs on midrib beneath ALASKA BLUEBERRY (*Vaccinium alaskaense*)

Alaska Blueberry

Vaccinium alaskaense T.J. Howell

DESCRIPTION Spreading to erect shrub to 6 feet (2 m) high.

LEAVES 3/4–2 in. (2–5 cm) long and 3/8–1 in. (1–2.5 cm) wide, thin, entire or shallowly toothed on edges, upper surface green, lower surface whitish (glaucous), with few short glandular hairs on midvein.

TWIGS thin, 1/32 in. (1–1.5 mm) in diameter, weakly angled, yellow green, becoming gray with age, ending in narrow stub. **Buds** green or red, 1/8 in. (3–4 mm) long, with 2 even scales, end bud lacking.

FLOWERS single at base of leaves after leaves are partially developed, on straight stalks 1/4–3/8 in. (6–10 mm) long; corolla bronzy pink, rounded urn-shaped, 1/4–5/16 in. (6–8 mm) long, widest just above base, usually broader than long. FRUIT a berry, bluish black to purple, variable in shape, usually without a bloom, 5/16–5/8 in. (8–15 mm) in diameter, on a stalk often more than 3/8 in. (1 cm) long, straight, or nearly so, enlarged just below fruit.

FLOWERING in April and May, berries ripening from mid-July to mid-August.
HABITAT Common in the spruce-hemlock forests of the coast (especially in forest openings and on cutover land).

USES The berries of this species and of early blueberry are usually picked together, as the shrubs occur in similar habitats. Used widely for jam and jelly and frozen for winter use. The berries are eaten by bears; the twigs are browsed by goat, elk, and deer.

NOTE For differences between Alaska blueberry (*Vaccinium alaskaense*) and the similar early blueberry (*V. ovalifolium*), see the latter.

ALASKA BLUEBERRY

ALASKA BLUEBERRY

Dwarf Blueberry
Vaccinium cespitosum Michx.

OTHER NAMES swamp blueberry, dwarf bilberry, dwarf huckleberry

DESCRIPTION Low spreading shrub forming mats to 16 in. (40 cm) high.

LEAVES elliptic to oblanceolate, 3/8–1 in. (10–25 mm) long and 3/16–3/8 in. (5–10 mm) wide, rounded to short-pointed at tip, edges with fine teeth usually gland-tipped, net-like veins conspicuous in some leaves but obscure on others; upper surface green, lower surface lighter; both hairless or with scattered short stiff hairs.

TWIGS much branched, often rooting at nodes, young twigs slender, green, with short hairs, round or sometimes angled, **older twigs** brown to gray, bark usually shredding. **Buds** small, red or green with 2 even scales, which meet at the edge.

FLOWERS single, at base of leaves, nodding on stalks 1/8 in. (3 mm) long, corolla white or pink, urn-shaped 1/4–5/16 in. (6–8 mm) long, with 5 small rolled lobes.

FRUIT a blue berry 1/4–5/16 in. (6–8 mm) in diameter, with a bluish bloom, sweet.

FLOWERING from late May through mid-July, fruit ripening in August.

HABITAT Common shrub of bogs, subalpine meadows, and open spruce-hemlock stands in the coastal forest and is occasional in white spruce and paper birch stands in the southern parts of the boreal forest. It also occurs above tree-line in the coastal mountains to elevations of 3,800 feet. From the south slopes of the Alaska Range to the Kenai Peninsula and southward throughout all of southeastern Alaska.

USES The fruits, which ripen early in August, are eaten raw or made into jams and jellies.

DWARF BLUEBERRY

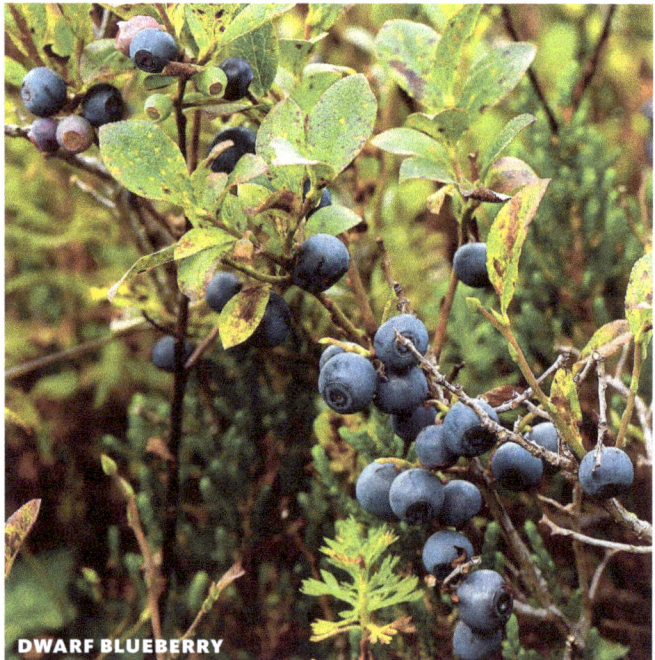

DWARF BLUEBERRY

Early Blueberry
Vaccinium ovalifolium Sm.

OTHER NAMES blue huckleberry, ovalleaf whortleberry, blue whortleberry

DESCRIPTION Early flowering, spreading shrub to 5 feet (1.5 m) tall.

LEAVES oval, rounded at tip and base, 3/4–2 in. (2–5 cm) long and 3/8–1 in. (1–2.5 cm) wide, thin, entire to shallowly toothed on edges, hairless, green on upper surfaces, whitish (glaucous) beneath, leaves at tips of twigs usually largest.

TWIGS slender, 1/16 in. (1.5–2 mm) in diameter, yellowish green to reddish, shiny, weakly angled, becoming gray the 2nd or 3rd year, ending in narrow stub. **Buds** green or red, 1/8 in. (3 mm) long, with 2 even scales, end bud lacking.

FLOWERS in spring before the leaves, single on sides of twig, nodding on stalks 3/16–1/4 in. (5–6 mm) long; corolla pink, urn-shaped, 3/16–5/16 (5–8 mm) long, broadest below the middle and usually longer than broad.

FRUIT a round blue to bluish black berry, with a bluish bloom; stalk usually less than 3/8 in. (1 cm) long, curved, not enlarged below fruit.

FLOWERING in April and May, berries ripening in mid-July to August.

HABITAT The most common blueberry of the coastal forest, where it may form a nearly continuous shrub layer under an open tree canopy and on cutover forest land.

USES This species and Alaska blueberry (*Vaccinium alaskaense*) provide most of the blueberries picked in coastal Alaska where they are made into jellies and jams and frozen for use in winter. The shrub is also used as winter browse by deer, mountain-goat, and elk.

EARLY BLUEBERRY

NOTE Early blueberry and Alaska blueberry are very similar in appearance; the following may help differentiate them:

Early Blueberry (*Vaccinium ovalifolium*)
- leaves hairless
- flowering before or with the leaves
- corolla usually longer than broad, pink, style included
- berry bluish or blue-black, with whitish bloom
- fruit stalks usually less than 3/8 in. (1 cm) long, curved, not enlarged just below the fruit

Alaska Blueberry (*Vaccinium alaskaense*)
- leaves with few short glandular hairs along midvein on lower surface
- flowering after the leaves are half developed
- corolla usually broader than long, bronzy pink, style exerted
- berry blue-black, without whitish bloom
- fruit stalks often more than 3/8 in. (1 cm) long, straight or nearly so, somewhat enlarged just below fruit

EARLY BLUEBERRY

EARLY BLUEBERRY - FLOWERS

Bog Cranberry
Vaccinium oxycoccos L.

OTHER NAMES swamp cranberry, wild cranberry, small cranberry

SYNONYM *Oxycoccus microcarpus* Turcz.

DESCRIPTION Evergreen shrub with very slender stems, creeping vinelike through moss and rooting at nodes.

LEAVES persistent, small, lance-shaped, 1/8–3/8 in. (3–10 mm) long, 1/32–1/8 in. (1–3 mm) wide, short-pointed, leathery, edges strongly rolled under; shiny dark green on upper surface, gray or whitish beneath with conspicuous midrib.

STEMS yellow to reddish brown, trailing, very slender, 1/32–1/16 in. (1–1.5 mm) in diameter, hairless when young.

FLOWERS 1–4 at ends of stems, nodding on erect slender stalks 3/4–1½ in. (2–4 cm) long with 2 tiny bractlets below middle; petals 4, red to pink, bent backward, 1/4 in. (6 mm) long; 8 stamens 1/8 in. (3 mm) long, yellow, pointing forward.

FRUIT a red, juicy, round berry 1/4–3/8 in. (6–10 mm) in diameter.

FLOWERING in June, berries ripening in August.

HABITAT Found in most sphagnum bogs and peat hummocks in the coastal and boreal forests but seldom abundant. Ranging from coastal forests north to south slopes of Brooks Range, westward to Bering Strait and tip of Aleutians (though absent from many islands).

USES The berries are good tasting and can be eaten raw or prepared as jelly or jam in the same manner as the closely related commercial cranberries. However, the bog cranberry seldom is abundant enough to be gathered in large quantities.

NOTE As the petals are bent backward, the cranberry flower resembles that of a miniature shootingstar. Because it is so tiny, the plant is often overlooked until the berries turn red in the fall.

BOG CRANBERRY

BOG CRANBERRY - FRUIT

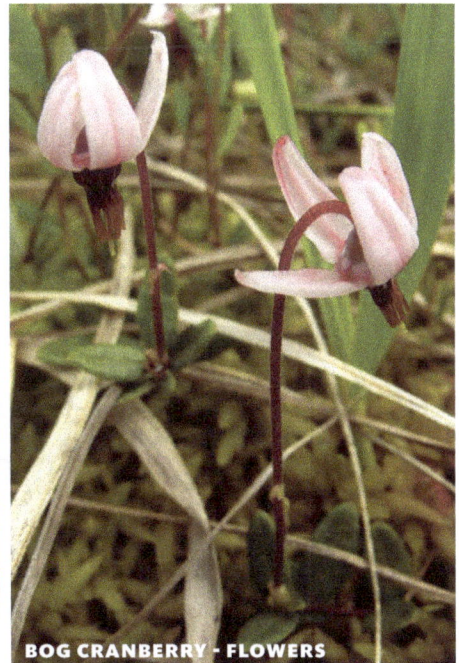

BOG CRANBERRY - FLOWERS

Red Huckleberry
Vaccinium parvifolium Sm.

OTHER NAME red whortleberry

DESCRIPTION Erect shrub 3–10 feet (1–3 m) tall, with small leaves and red berries.

LEAVES deciduous, often persisting on twigs into early winter, oval to elliptic, 3/8–1¼ in. (1–3 cm) long and 1/4–3/8 in. (6–10 mm) wide, entire, green on upper surface and grayish beneath; petioles short, 1/32 in. (1 mm) long.

TWIGS slender, green, shiny, strongly angled or ridged, ending in narrow stub. Buds light green, 1/8–3/16 in. (3–5 mm) long, covered by 2 scales, end bud lacking.

FLOWERS single at base of leaves on stalks 1/4–3/8 in. (6–10 mm), nodding, corolla broadly urn-shaped with 5 small lobes, waxy, yellowish pink to red, 1/8–1/4 in. (3–6 mm) long.

FRUIT a bright red round berry, 1/4–3/8 in. (6–10 mm) in diameter.

FLOWERING in May and June, berries ripening mid-to late August.

HABITAT An occasional to common shrub in openings along roadsides, and in cutover forest land, in the coastal spruce-hemlock forests; southeastern Alaska north to Yakutat Bay.

USES The berries are sour but with good flavor and are used for jelly. The green twigs are commonly browsed by deer, elk, and goats in fall and winter, and the berries are eaten by blue grouse and bears.

RED HUCKLEBERRY

RED HUCKLEBERRY - FLOWERS

RED HUCKLEBERRY - FRUIT

Bog Blueberry
Vaccinium uliginosum L.

OTHER NAMES alpine blueberry, bog bilberry, great bilberry, whortleberry

DESCRIPTION Much branched low shrub, erect or prostrate, 8–16 in. (20–40 cm) high, often rooting along branches.

LEAVES oval (obovate) to elliptic, 3/8–3/4 in. (1–2 cm) long, dark green on upper surface, lighter below, with conspicuous veins.

TWIGS slender, round, 1/32 in. (1 mm) in diameter, brown, minutely hairy, older twigs much branched, yellow-brown to gray with shredding bark. **Buds** small, 1/32 in. (1 mm) long, several scales with scattered short hairs.

FLOWERS 1–4 from ends or side branches, nodding on stalk 1/16–1/8 in. (1.5–3 mm) long; corolla pink, urn-shaped, 1/8–3/16 in. (3–5 mm) long with 4 short lobes.

FRUIT a blue to black berry with bluish bloom, ovoid 3/16–3/8 in. (5–10 mm) in diameter.

FLOWERING in June, berries ripening in late July and August.

HABITAT A very common shrub of bog, open forest, and tundra of all of Alaska except for the extreme northern coastal plain. In southeastern Alaska it grows in the alpine tundra to elevations of 5,600 feet.

USES The berries are picked in large quantities in north, central, and western Alaska, but not used extensively in southeastern Alaska where other blueberries are more readily available. The berries are also eaten by bears, grouse, and ptarmigan. Blueberries of this and related species are eaten raw or cooked in pies, puddings, and muffins, and may be frozen or canned. They are a fair source of vitamin C.

BOG BLUEBERRY

BOG BLUEBERRY

Mountain-Cranberry
Vaccinium vitis-idaea L.

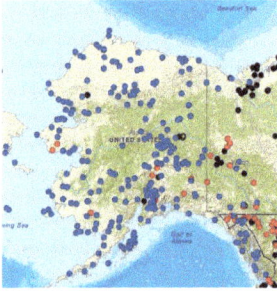

OTHER NAMES lingenberry, lingberry, lowbush cranberry, partridgeberry, cowberry

DESCRIPTION Evergreen creeping, mat-forming subshrub 2–6 in. (5–15 cm) tall, with shiny leaves and bright red berries.

LEAVES oval, 3/8–3/4 in. (10–20 mm) long and 1/4–3/8 in. (6–10 mm) wide, thick, green, and shiny above, light green beneath and spotted with short stiff brown hairs, edges slightly rolled under.

STEMS slender and trailing, rooting at nodes, light brown to yellow.

FLOWERS 1 to several, nodding on short stalks 1/16 in. (1–2 mm) long at ends of twigs, corolla pink, bell-shaped, about 3/16 in. (5 mm) long, with 4 short lobes.

FRUIT a bright red, sour berry, 1/4–5/16 in. (6–8 mm) in diameter.

FLOWERING in mid-to late June and July, berries ripening in August.

HABITAT common in spruce and birch woods of the boreal forest, in bogs and alpine types in most of Alaska, and in the tundra of the north and western regions. It usually forms a loose mat in moist mossy situations but also forms dense mats in dry rocky slopes in arctic and alpine areas.

USES The berries are abundant and usually picked in the fall after the first frost but may remain under the snow during the winter and become available in the spring when the snow melts. They are commonly used for jams, jellies, relishes, and beverages. Although sour, they have a better flavor than the commercial cranberry. The berries also provide a source of food for ptarmigan, grouse, and bears. The foliage is reported to be of some value as winter browse for reindeer and caribou.

NOTE North American plants are smaller in leaf and berry size than those in the Old World and have been named subspecies *Vaccinium vitis-idaea* subsp. *minus* (Lodd.) Hultén.

MOUNTAIN-CRANBERRY

MOUNTAIN-CRANBERRY

MOUNTAIN-CRANBERRY

■ GOOSEBERRY FAMILY *Grossulariaceae*

Shrubs with erect, spreading, or prostrate branches. **Leaves** alternate, palmately veined and palmately lobed, frequently with glandular hairs. **Twigs** with or without prickles and spines, angled, with papery shedding bark; pith porous or spongy. **Flowers** usually in racemes, but occasionally solitary, borne on side shoots with 1 or 2 leaves at base, small; tubular base with 5 sepals larger and more conspicuous than the 5 scalelike petals; stamens 5; pistil with inferior 1-celled ovary and 2 styles. **Fruit** a many-seeded berry with dried remains of flower at tip.

CURRANT, GOOSEBERRY *Ribes*

One genus, *Ribes,* occurs in Alaska, containing both gooseberries (one species) and currants (5 species). Species with spines or prickles on their stems are usually called gooseberries, and those with unarmed branches, currants. Both groups are used for making jams and jellies.

KEY TO ALASKA *RIBES*

1 Stems armed with spines and prickles; leaves small, less than 2 in. (5 cm) long

SWAMP GOOSEBERRY (*Ribes lacustre*)

1 Stems unarmed; leaves larger, more than 2 in. (5 cm) long   2

2 Ovary and fruit with resin dots   3

2 Ovary and fruit without resin dots, often with stalked glands   4

3 Racemes 6–12 in. (15–30 cm) long, with 20–40 flowers; sepals greenish, fruit with white to bluish bloom; twigs coarse, 1/8–1/4 in. (3–6 mm) in diameter, brownish, shedding bark; leaves longer than broad, underside with resin glands STINK CURRANT (*Ribes bracteosum*)

3 Racemes 3 in. (8 cm) long, 6–12 flowered; sepals whitish; fruit black without bluish bloom; twigs slender, 1/4 in. (6 mm) or less in diameter, gray with black spots, smooth; leaves broader than long, underside without resin glands

NORTHERN BLACK CURRANT (*Ribes hudsonianum*)

4 Berries smooth, without stalked glands, red; sepals reddish; flower racemes drooping, leaves mostly 3-lobed, occasionally with pair of smaller lobes at base, not divided to middle   AMERICAN RED CURRANT (*Ribes triste*)

4 Berries with stalked glands, red or black to dark blue; sepals green, white, or light pink; flower racemes ascending; leaves 5-lobed, divided to middle   5

5 Berries red, sepals white to pink, without hairs; twigs fine, less than 1/8 in. (3 mm) in diameter   SKUNK CURRANT (*Ribes glandulosum*)

5 Berries black to dark blue; sepals green to white, with hairs on back; twigs coarse, more than 1/8 in. (3 mm) in diameter   TRAILING BLACK CURRANT (*Ribes laxiflorum*)

Stink Currant
Ribes bracteosum Dougl. ex Hook.

OTHER NAME blue currant

DESCRIPTION Erect to spreading shrub, 3–8 feet (1–2.5 m) tall, with large leaves and long racemes of flowers and fruits, and sweet, rather disagreeable odor.

LEAF BLADES 3–8 in (7.5–20 cm) long and slightly broader, 5–7 lobed, lobes toothed at edge and short-pointed at tip, underside dotted with tiny resin glands. Petioles variable in length, from shorter to much longer than blade.

TWIGS coarse, those of previous year to 1/4 in. (6 mm) in diameter, brown to grayish, often with shredded bark; **buds** large and red.

FLOWERS in long erect to ascending racemes 6–12 in. (15–30 cm) long with 20–40 flowers; stalks slender, to 3/8 in. (1 cm) long, with leaflike bract at base often exceeding the stalk, sepals white or greenish, often with purple tinge, spreading, 1/8 in. (3–4 mm) long; ovary conspicuously glandular.

FRUIT a spherical berry 3/8 in. (1 cm) in diameter, glandular, with white to bluish bloom and fetid odor.

FLOWERING in May and June, fruit ripening in late July and August.

HABITAT Commonly with alder in openings in coastal spruce hemlock forests, and in disturbed wet places along roadsides.

USES In spite of the strong odor, Alaska Natives along the coast consume the fruits after mixing with salmon roe and storing for the winter.

STINK CURRANT

STINK CURRANT

Skunk Currant

Ribes glandulosum Grauer

OTHER NAME fetid currant

SYNONYM *Ribes prostratum* L'Her.

DESCRIPTION Low shrub 2–3 feet (0.6–0.9 m) high with sprawling or reclining branches and strong fetid odor.

LEAF BLADES 1–3 in. (2.5–7 cm) long and slightly broader, divided nearly to middle into 5 lobes, the lower pair smaller. Petioles about equal to blade, with a few bristle-like hairs near base.

TWIGS smooth and grayish, becoming brown and with shredded bark with age. **Buds** 1/4–3/8 in. (6–10 mm) long, reddish, with fine white hairs at tip of scales.

FLOWERS 10–20 in erect racemes, 3–4 in. (7.5–10 cm) long that droop when the fruit ripens, individual flower and fruit stalks (pedicels) 3/16–5/16 in. (5–8 mm) long with gland-tipped hairs; sepals spreading, white to pinkish, rounded, 1/16 in. (2 mm) long; ovary with gland-tipped hairs.

FRUITS bristly red berries 1/4 in. (6 mm) in diameter, with gland-tipped hairs.

FLOWERING in June, berries ripening in late July and early August.

HABITAT Most commonly found in disturbed areas beside roads and adjacent to cleared fields.

NOTE In spite of its strong odor, it makes excellent jelly.

SKUNK CURRANT

SKUNK CURRANT

Northern Black Currant
Ribes hudsonianum Richards.

OTHER NAME Hudson Bay currant

DESCRIPTION Usually an erect shrub 3–6 feet (1–2 m) tall but northward often prostrate and spreading, 1–3 feet (3–9 dm) tall, with strong, rather unpleasant odor when leaves or berries are crushed.

LEAF BLADES 3–4 in. (8–10 cm) wide and 2–3 in. (5–8 cm) long, broadly 3-lobed about 1/3 to midvein, lobes sharply toothed at edge, with resin dots and scattered hairs on lower surface; petioles about 2/3 as long as blade.

TWIGS gray and shiny, scattered with small black glands or short black hairs. **Buds** red, 1/8–1/4 in. (3–6 mm) long hairless, on short stalks.

FLOWERS 6–12 in short racemes 2–3 in. (5–8 cm) long; sepals white, triangular, elongate, 1/8–3/16 in. (3–5 mm) long; ovary resin dotted.

FRUIT an oval berry, black, usually with resin dots but without bloom, bitter.

FLOWERING in June and July, fruits ripening in July and August.

HABITAT A common shrub of boreal forests in spruce, birch, and aspen forest types. Near tree-line it grows with alders. Primarily an interior species but reaching the coast at Seward, Prince William Sound, and in the vicinity of Haines and Juneau; also reported from southeastern Alaska. North to the Brooks Range, and west to lower Yukon and Kuskokwim Rivers.

NOTE The berries are not eaten because of their bitter taste.

NORTHERN BLACK CURRANT

NORTHERN BLACK CURRANT

Swamp Gooseberry
Ribes lacustre (Pers.) Poir.

OTHER NAMES prickly currant, swamp currant, swamp black currant, bristly black currant

SYNONYMS *Ribes oxycanthoides* var. *lacustre* Pers.

DESCRIPTION Usually a spreading shrub, sometimes erect, 2–4 feet (3–12 dm) tall, with spiny twigs and deeply dissected leaves with skunklike odor.

LEAF BLADES 1½–2 in. (3.5–5 cm) long and 1½–2 in. (4–5 cm) wide, 5-lobed and divided 2/3–3/4 to midrib, the lower pair of lobes smaller, each lobe again dissected into several rounded teeth. Petioles 3/4–1½ in. (2–4 cm) long, with bristly hairs.

TWIGS yellowish brown, densely to sparsely covered with sharp spines, 1/8–3/16 in. (3–5 mm) long with a few larger spines at nodes.

FLOWERS 6–15 on a drooping raceme, sepals light green to purplish, oval, 1/8 in. (2.5–3 mm) long, covered with gland-tipped hairs.

FRUIT a berry 1/4–5/16 in. (6–8 mm) in diameter, black to dark purple, bristly with gland tipped hairs.

FLOWERING in June, fruit ripening in August.

HABITAT An occasional shrub with white spruce and Sitka spruce in the interior and coastal forests. Because of the occurrence in isolated clumps, commonly low production, and the skunklike odor, the bristly berries are not often used for making jellies and jams.

ADDITIONAL SPECIES Canada gooseberry (*Ribes oxycanthoides* L.) has been recorded from several localities in south-central Alaska. This spiny shrub resembles swamp gooseberry somewhat, but has flowers and fruits single or paired along the stem. Its berries are edible but sour.

SWAMP GOOSEBERRY

SWAMP GOOSEBERRY

SWAMP GOOSEBERRY

Trailing Black Currant
Ribes laxiflorum Pursh

DESCRIPTION Usually a low spreading shrub with branches running along the ground, but sometimes vinelike and climbing on erect shrubs. **LEAF BLADES** $2\frac{1}{2}$–3 in. (6–8 cm) long and 3–4 in. (7–10 cm) across, divided into 5 deep, triangular lobes doubly toothed along edge with sharp or rounded teeth, lower surface light green with small yellow glands near base. Petiole 2–3 in. (5–7.5 cm) long.

TWIGS yellow brown and hairy when young, stout, 1/8–3/16 (3–5 mm) in diameter, becoming dark brown and slightly fissured. **Buds** 1/4–5/16 in. (6–8 mm) long, light to dark red, hairy on surface and edges.

FLOWERS 10–20 in a raceme 4–6 in. (10–14 cm) long; sepals 1/8 in. (3 mm) long, greenish white,

TRAILING BLACK CURRANT

red, or dark purple, broadly triangular and rounded at tip; with gland-tipped hairs on ovary and pedicel 1/4 in. (6–8 mm) long.

FRUIT a black berry 1/2–5/8 in. (12–15 mm) in diameter with bluish bloom and gland-tipped hairs on surface, with fetid odor when crushed.

FLOWERING in early to late May at the time of leafing, fruits ripening in late July to early August.

HABITAT primarily a low, spreading shrub of disturbed ground, open meadows, cut-over forest land, and dense spruce-hemlock forests of coastal Alaska.

NOTE In Oregon and Washington, this shrub may become vinelike and reach heights of 20 feet (6 m), but in Alaska seldom more than 4 feet (1.2 m) high.

TRAILING BLACK CURRANT

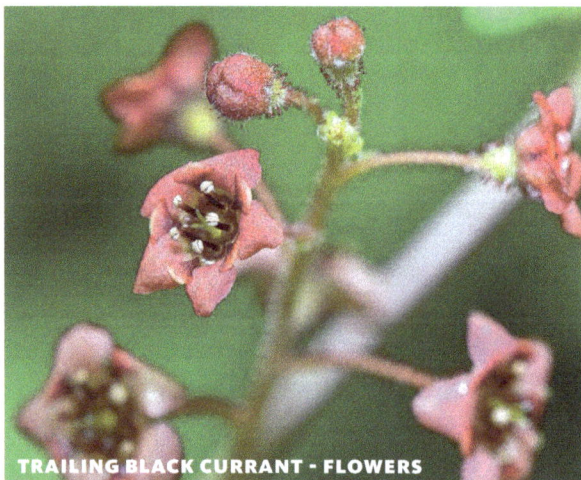

TRAILING BLACK CURRANT - FLOWERS

American Red Currant
Ribes triste Pall.

OTHER NAME northern red currant

DESCRIPTION Low spreading shrub with bright red berries, branches prostrate and frequently rooting at nodes, sometimes erect to 2–3 feet (6–9 dm) high.

LEAF BLADES 4 in. (10 cm) long and 2–3 in. (5–8 cm) broad but along coast becoming somewhat larger (10 x 4 in. or 25 x 10 cm), usually 3-lobed but often with pair of small lobes near base, lobes broadly triangular and coarsely toothed along edges. Petiole 1/2–2/3 as long as blade.

YOUNG TWIGS smooth and light brown but soon becoming shredded and reddish brown, a characteristic feature in winter. **Buds** dark red, 3/16–1/4 in. (5–6 mm) long.

FLOWERS 6–20 on a drooping raceme 2–4 in. (5–10 cm) long; sepals rounded, 1/16 in. (2 mm) long, purplish, inconspicuous.

FRUIT a translucent red berry 1/4–3/8 in. (6–10 mm) in diameter, smooth, sour.

FLOWERING in May and early June before or with the leaves, fruit ripening in August.

HABITAT A rather common shrub in the white spruce and paper birch forests of the interior of Alaska. North and west of the tree-line, it is found in willow and alder thickets in protected ravines. In southeast Alaska, it grows only at the heads of some of the fiords, usually in alder thickets.

NOTE American red currant closely resembles the commercially grown currants and is widely used in Alaska for jellies and jams as well as eaten raw.

AMERICAN RED CURRANT

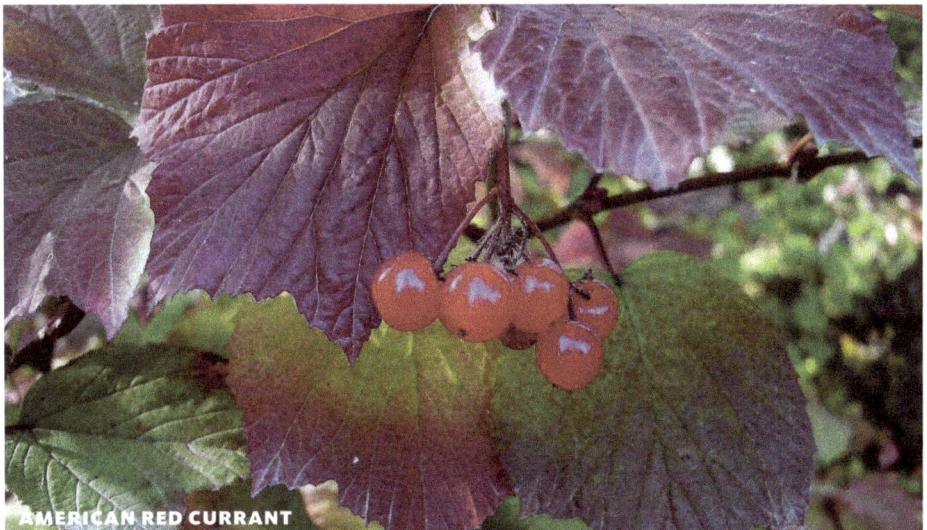

AMERICAN RED CURRANT

■ TWINFLOWER FAMILY *Linnaeaceae*

Twinflower
Linnaea borealis L.

SYNONYM *Linnaea americana* Forbes

DESCRIPTION Creeping evergreen dwarf shrub or herbaceous, forming loose mats, with long slender, slightly hairy, woody horizontal stems, rooting at nodes, and many erect twigs to 4 in. (10 cm) high.

LEAVES opposite, with slender petioles less than 1/8 in. (3 mm) long, with stipules.

FLOWERS paired at tip of very slender erect stalks 1½ –3 in. (4–7.5 cm) long, nodding, fragrant, 3/8–5/8 in. (10–15 mm) long, composed of calyx of 5 narrow greenish hairy lobes, pink to purple funnel-shaped or bell-shaped tubular corolla with 5 nearly equal lobes, 4 stamens in pairs inserted near base of tube and enclosed, and pistil with inferior greenish 3-celled ovary, 1 ovule, and slender style.

FRUIT small, dry, round, 1/16 in. (1.5 mm) in diameter, with calyx at tip and enclosed by bracts, 1–seeded.

FLOWERING June–August, fruits maturing July–August.

HABITAT Scattered in open forests and tundra. Widespread almost throughout Alaska (except Arctic coastal plain), from Aleutian Islands through interior to southeast Alaska.

NOTE Twinflower can be transplanted into cultivation as a spreading ever-green ground cover in shady places.

ETYMOLOGY The generic name hon-ors Carolus Linnaeus (1707–78), Swedish botanist considered the father of modern plant taxonomy.

TWINFLOWER

TWINFLOWER

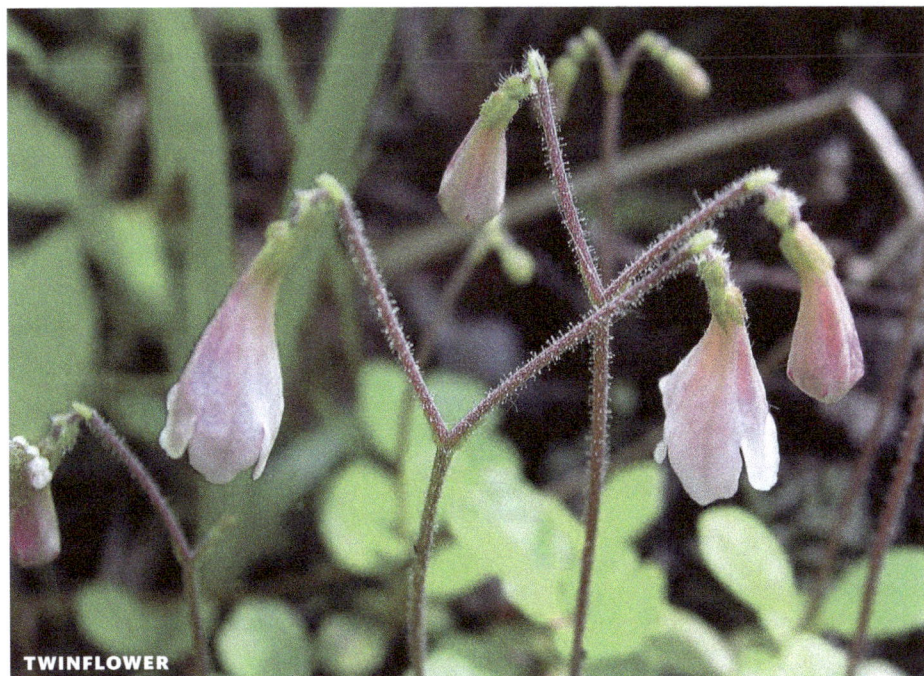

TWINFLOWER

■ BAYBERRY FAMILY *Myricaceae*

Fragrant shrubs in Alaska (elsewhere also small trees). **Leaves** alternate, simple with entire or toothed margins, and with small aromatic yellow resin dots on both surfaces. **Flowers** small, yellowish, without sepals or petals, unisexual, male and female on separate plants in short scaly spikes (aments), stamens usually 4–8, ovary 1-celled. **Fruit** tiny, nutlike, with whitish waxy coat, 1–seeded. Family includes bayberry (*Myrica*) and sweetfern (*Comptonia*) of eastern United States. In Alaska only one species.

Sweetgale
Myrica gale L.

DESCRIPTION Low shrub 1–4 feet (3–12 dm) tall, of low wet habitats, usually branching loosely at base.

LEAVES oblanceolate, 1–2 in. (25–50 mm) long and 3/8–1/2 in. (10–12 mm) wide, rounded at tip with several coarse teeth, tapering at base to short petiole 1/8 in. (3–5 mm) long, thinly hairy on both surfaces and dotted with yellow waxy glands.

TWIGS slender, finely hairy when young, dark brown to gray with yellow resin dots and white dots (lenticels), resembling birch and alder. **Buds** 1/8 in. (3 mm) long, pointed, dark reddish brown, hairless.

FLOWERS male and female on separate plants (dioecious), small, inconspicuous, yellowish, in spikes in early spring before the leaves. **Male (staminate) spikes** 3/8–5/8 in. (10–15 mm) long, **female (pistillate) spikes** 1/4–3/8 in. (6–10 mm), both dotted with yellow waxy glands.

FRUIT a green 2–winged nutlet 1/8 in. (3 mm) long, resinous waxy.

FLOWERING from mid-May to the first week in June (depending on locality); sweetgale is one of Alaska's earliest blooming plants.

HABITAT A common shrub of low wet areas, especially bogs in interior Alaska and tidal flats along the coast.

NOTE The following year's flower spikes form in late summer and the stalks of previous summer's spikes often remain throughout the winter, giving the winter twigs a distinctive appearance.

SWEETGALE

SWEETGALE

SWEETGALE - FEMALE SPIKES

■ ROSE FAMILY *Rosaceae*

Woody plants in the rose family (Rosaceae) are well-represented in Alaska by ca. 25 species and 12 genera. Distinguishing characters are:

- **leaves** alternate, simple or pinnately or palmately compound, with paired stipules;
- **flowers** regular, often large and showy or small and many, with cuplike base, 5 sepals mostly persistent, 5 petals mostly white or less commonly pink, purple, or yellow, many stamens, and usually 1 pistil with 2–5 celled ovary (often inferior) and 2–5 styles (sometimes many simple pistils); and
- **fruit** variable, like an apple (pome) or plum (drupe), aggregate of many 1-seeded fruits ("berry"), or many separate fruits.

Numerous wild and cultivated fruits and ornamental plants belong to this family. Several native genera produce fruits edible to humans as well as wildlife, for example, serviceberry (*Amelanchier*), crab apple (*Malus*), raspberry (*Rubus*), and strawberry (*Fragaria*). Others, such as mountain-ash (*Sorbus*), mountain-avens (*Dryas*), and spiraea (*Spiraea*) are ornamentals. Rose (*Rosa*) is both ornamental and edible, rose hips being a good source of vitamin C.

Four native species of the rose family and one naturalized species become small trees in Alaska; these species are: Pacific serviceberry (*Amelanchier alnifolia*), Oregon crab apple (*Malus fusca*), Cascade mountain-ash (*Sorbus scopulina*), Sitka mountain-ash (*Sorbus sitchensis*), and the naturalized European mountain-ash (*Sorbus aucuparia*). In addition, two cherries (*Prunus*) are reported as uncommon escapes from cultivation.

SERVICEBERRY *Amelanchier*

Deciduous shrubs or small trees. **Leaves** alternate, simple, with paired narrow stipules soon shedding, short petioles, and mostly small elliptic to oblong blades mostly rounded at both ends and coarsely toothed on edges. **Buds** oblong or conical, long and narrow, covered by several overlapping scales. **Flowers** several in small terminal clusters (racemes), appearing with or before the leaves; calyx of 5 persistent lobes or sepals; petals 5, showy, narrow, white; stamens 10-20 ; pistil with inferior, 2–5-celled ovary, and 2–5 styles mostly united at base. **Fruit** like a small apple (pome), round, dark blue or purple, with calyx at tip, juicy and sweet, containing 4–10 seeds and cells.

Saskatoon Serviceberry
Amelanchier alnifolia (Nutt.) Nutt.

OTHER NAMES northwestern serviceberry, western serviceberry, juneberry

DESCRIPTION Shrub 3–6 feet (1–2 m) high (small tree to 16 feet (5 m) southward) and 5 in. (12.5 cm) in trunk diameter.

LEAVES with slender petioles 3/8–3/4 in. (1–2 cm) long, hairy when young. **Blades** nearly round varying to elliptic or oblong, 3/4–1½ in. (1–4 cm) long, 5/8–1¼ in. (1.5–3 cm) wide, rounded at both ends coarsely toothed above middle, thick and firm, above dark green and hairless, beneath paler and hairy when young.

FLOWER CLUSTERS (racemes) 1¼ –2½ in. (3–6 cm) long. **Flowers** 5–15, fragrant, 1/2–3/4 in. (1.2–2 cm) broad, composed of calyx of 5 narrow lobes, densely woolly when young, 5 white oblong petals 3/8–5/8 in. (10–15 mm) long, about 20 short stamens, and pistil with inferior hairy 5-celled ovary and 5 styles.

FRUIT like an apple (pome), rounded, 3/8–5/8 in. (1–1.5 cm) in diameter, purple or nearly black and covered with a bloom, sweet, juicy, and edible, containing several elliptic flattened brown seeds 3/16 in. (5 mm) long.

FLOWERING in June, maturing fruit in July–August.

HABITAT Forests and openings, Pacific coast of southeast and southern Alaska; also on steep, dry south-facing bluffs, usually with aspen and common juniper.

USES Serviceberry fruits are eaten fresh or prepared in puddings, pies, and muffins; the dried berries are used like raisins and currants. Birds are fond of the fruits, but are rarely abundant enough to be significant as a wildlife food. Suitable for ornamental planting in interior Alaska for the attractive white flowers and fruits, the plants spreading and forming thickets.

SASKATOON SERVICEBERRY

SASKATOON SERVICEBERRY - FLOWERS

SASKATOON SERVICEBERRY - FRUIT

Black Hawthorn
Crataegus douglasii Lindl.

DESCRIPTION Deciduous shrub reported from southeast Alaska and in the Prince William Sound area; southward becoming a small spreading tree 25–40 feet (7.5–12 m) high and 1½ feet (45 cm) d.b.h.

LEAVES alternate, with paired broad, toothed stipules, slender petioles 1/2–3/4 in. (1.2–2 cm) long, and obovate to ovate thin blades 1–3 in. (2.5–7.5 cm) long and 5/8–2 in. (1.5–5 cm) wide, broadest toward the short-pointed tip, base short-pointed sharply toothed and often slightly lobed, becoming hairless, above shiny dark green, paler beneath.

TWIGS slender hairless shiny reddish, often with straight or slightly curved stout red to gray spines 3/8–1 in. (1–2.5 cm) long.

BARK gray, smoothish.

FLOWER CLUSTERS (corymbs) terminal, broad with several flowers 1/2 in. (1.2 cm) across on slender stalks, composed of greenish base (hypanthium), 5 long-pointed sepals reddish at end, 5 white rounded spreading petals 1/4 in. (6 mm) long, 10–20 stamens, and pistil with inferior 2–5-celled ovary and 2–5 styles.

FRUITS like small apple (pome), many in drooping clusters on long stalks, rounded, 1/2 in. (12 mm) in diameter, shiny black with calyx persistent at tip, thick light yellow flesh, sweetish and mealy but somewhat insipid and usually 5 nutlets 1/4 in. (6 mm) long.

HABITAT Very local in southeastern Alaska.

ADDITIONAL SPECIES **Huckleberry hawthorn** (*Crataegus gaylussacia* A. Heller) is reported from southeast Alaska. It is distinguished from *C. douglasii* by having thorns 3/8–1/2 in. (9–12 mm) long, and flowers with about 20 stamens. In *C. douglasii,* the thorns are longer, 5/8–7/8 in. (15–23 mm), and the flowers have only about 10 stamens.

BLACK HAWTHORN

BLACK HAWTHORN

BLACK HAWTHORN

Bush Cinquefoil
Dasiphora fruticosa (L.) Rydb.

OTHER NAMES shrubby cinquefoil, yellow-rose
SYNONYM *Potentilla fruticosa* L.
DESCRIPTION Much branched deciduous shrub 1–5 1/2 feet (0.3–1.7 m) high.
LEAVES alternate, pinnate, ¾–1½ in. (2–3.5 cm) long, with paired clasping, ovate, light brown, hairy, persistent stipules 1/4–1/2 in. (6–12 mm) long, with very slen-

der light brown hairy axis. **Leaflets** 5, stalkless, close together near end of axis and paired except at end, narrowly oblong or oblanceolate, 1/4–3/4 in. (6–20 mm) long and 1/16–1/4 in. (2–6 mm) wide, short-pointed at both ends, edges turned under, above dull green with inconspicuous pressed hairs, beneath whitish green with silky hairs.
TWIGS slender, light brown, with long silky hairs, becoming hairless.
BARK brown gray, shreddy.
FLOWERS borne singly at leaf bases or 3–7 in small terminal clusters (cymes), erect on slender silky hairy stalks, large and

BUSH CINQUEFOIL

showy ¾–1¼ in. (2–3 cm) across, composed of saucer-shaped hairy base (hypanthium), 5 narrow green bracts 1/4 in. (6 mm) long, 5 spreading ovate hairy sepals 1/4 in. (6 mm) long, 5 rounded spreading yellow petals 3/8–5/8 in. (10–15 mm) long, 20–30 short stamens, and many pistils with very hairy 1-celled ovary, 1 ovule, and short persistent style attached on side.

FRUITS (achenes) many, egg-shaped, 1/16 in. (2 mm) long, light brown and covered with whitish hairs, 1-seeded.

FLOWERING June–August, fruits maturing July–September and persistent.

HABITAT Common shrub in moist soil of swamps and borders of streams and lakes; also on dry rocky hillsides. Almost throughout Alaska except western Alaska Peninsula, Aleutian Islands, and most of southeast.

NOTE Wild plants tested in interior Alaska as ornamentals have scraggly growth. Several horticultural varieties including dwarf and large-flowered are cultivated elsewhere. It is reported that the leaves have been used for tea by the Eskimos at Nome.

BUSH CINQUEFOIL

MOUNTAIN-AVENS *Dryas*

Evergreen densely tufted, herbaceous dwarf shrubs with prostrate stems woody at base, branching, rooting, often forming large rounded mats or clumps. **Leaves** crowded but alternate, with 2 narrow long-pointed stipules attached to slender petiole. Blade mostly oblong, leathery, with wavy toothed or straight edges, dark green above and densely white-hairy beneath. **Flowers** many and showy, solitary on erect stalks, 3/4–1 in. (2–2.5 cm) across, composed of saucerlike or convex base (hypanthium), calyx of 8–10 persistent sepals, 8–10 widely spreading white petals (pale yellow and slightly spreading in *Dryas drummondii*), many stamens, and many pistils, with 1-celled ovary, and slender hairy styles forming feathery plumes.

All members of the genus, but especially *Dryas drummondii,* are associated with calcareous rocks or soil. Therefore, the presence of *Dryas* on non-calcareous bedrock (or on soils derived from such rocks), indicates the presence of local calcium-rich mineral veins, or of glacier- or water-transported calcareous soil.

Besides the 3 species generally accepted and illustrated here, variations and hybrids have been described, and additional species or subdivisions listed under other names are sometimes reported for Alaska, especially for *Dryas octopetala.* The key separates Alaska's three most common and well-defined *Dryas* species.

KEY TO ALASKA *DRYAS*

1 Leaf base short-pointed (wedge-shaped); flowers nodding with pale yellow, slightly spreading petals   **DRUMMOND MOUNTAIN-AVENS** (*Dryas drummondii*)
1 Leaf base straight or notched (heart-shaped); flowers erect with white, widely spreading petals   **2**
2 Leaves with margins wavy-toothed from tip to base, very rough on upper surface, with glands and scales on midvein beneath   **EIGHT-PETAL MOUNTAIN-AVENS** (*Dryas octopetala*)
2 Leaves with margins straight (entire) or slightly wavy in lower half, not rough or slightly rough on upper surface, without glands and scales on midvein beneath
     ENTIRE-LEAF MOUNTAIN-AVENS (*Dryas integrifolia*)

Drummond Mountain-Avens

Dryas drummondii Richards. ex Hook.

OTHER NAME yellow dryas

DESCRIPTION Evergreen herbaceous dwarf shrub with prostrate stems, forming large mats.

LEAVES crowded, with long slender petiole. **Blades** elliptic, 5/8–1¼ in. (1.5–3 cm) long, 3/8–3/4 in. (1–2 cm) wide, thick, rounded at tip and short-pointed at base, edges wavy-toothed and turned under, above dark green and usually slightly hairy with sunken veins, beneath densely white hairy.

FLOWERS solitary, nodding on whitish hairy stalks 2–8 in. (5–20 cm) high; yellow, 3/4–1 in. (2–2.5 cm) across; sepals 8–10, ovate, blackish, glandular-hairy, nearly 1/4 in. (6 mm) long; petals 8–10, yellow, nearly 1/2 in. (12 mm) long.

FRUITS headlike of many achenes 3/16 in. (4 mm) long with persistent long hairy styles, forming feathery plumes 1–1½ in. (2.5–3.5 cm) long, in a mass 1–2½ in. (2.5–6 cm) in diameter.

FLOWERING June–July, fruits maturing July–August.

HABITAT Arctic-alpine to lowland areas especially as a pioneer on gravel bars of floodplains. Interior Alaska in Brooks Range and from Alaska Range south to Kenai Peninsula and southeast Alaska.

ETYMOLOGY The scientific name honors the discoverer, Thomas Drummond (1780–1835), Scotch botanical explorer in North America.

DRUMMOND MOUNTAIN-AVENS

DRUMMOND MOUNTAIN-AVENS

Entire-Leaf Mountain-Avens
Dryas integrifolia Vahl

SYNONYMS *Dryas chamissonis* Spreng. ex Juz., *Dryas sylvatica* (Hultén) Porsild

DESCRIPTION Evergreen herbaceous dwarf shrub with prostrate stems.

LEAVES crowded with slender hairy petiole 1/4 in. (6 mm) long. **Blades** narrowly oblong or lanceolate, 3/8–1 (1½) in. (1–2.5 (3.5) cm) long, 1/8–3/8 in. (0.3–1 cm) wide, thick with blunt tip, broadest near rounded or notched base, edges mostly turned under and without teeth or with few wavy teeth toward base, above shiny dark green, smooth, and usually hairless, beneath densely white hairy.

FLOWERS solitary on erect stalks 1–4 (8) in. (2.5–10 (20) cm) high, hairy, and usually with blackish gland hairs; 3/4–1 in. (2–2.5 cm) across; sepals 8–9, narrow, glandular-hairy, nearly 1/4 in. (5 mm) long; petals 8–9, white, spreading, 3/8–1/2 in. (10–12 mm) long.

FRUITS headlike, of many achenes 1/8 in. (3 mm) long with persistent long hairy styles, forming whitish feathery twisted plumes 3/4–1¼ in. (2–3 cm) long, in a mass 1–1½ in. (2.5–3.5 cm) in diameter.

DRUMMOND MOUNTAIN-AVENS

FLOWERING May–August, fruits maturing June–August.

HABITAT Common and widespread in lowland and alpine tundra, on gravel bars and rocky slopes, in muskegs, also in open spruce stands near timberline. Arctic-alpine areas nearly throughout Alaska, from Bering Strait to Canadian border but not found in the extreme southeast.

NOTE A variation at low altitudes in interior Alaska, *Dryas integrifolia* subsp. *sylvatica* Hultén, has leaves long-stalked, long, thin, flat, with round tip, base rounded or short-pointed, and edges mostly without teeth, and taller flower stalks. It is found in bogs and spruce forests on gravel and limestone in interior Alaska (except southwest and southeast).

ENTIRE-LEAF MTN-AVENS

ENTIRE-LEAF MOUNTAIN-AVENS

ENTIRE-LEAF MOUNTAIN-AVENS

ENTIRE-LEAF MOUNTAIN-AVENS

Eight-Petal Mountain-Avens
Dryas octopetala L.

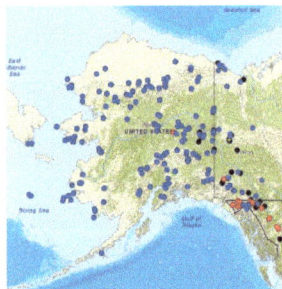

NOTE The description below encompasses plants traditionally classified as *Dryas octopetala* L.; however some authorities now consider that species to be present only in Greenland. As *Dryas octopetala* continues to be widely used in North America, that name is retained here.

SYNONYMS *Dryas ajanensis* Juz., *Dryas alaskensis* Porsild, *Dryas hookeriana* Juz.

DESCRIPTION Evergreen tufted herbaceous dwarf shrub with prostrate stems.

LEAVES crowded, with slender petiole. **Blades** oblong, 3/8–1¼ in. (1–3 cm) long, 3/16–3/8 in. (0.5–1 cm) wide, rounded at tip and short-pointed, rounded, or notched at base, edges coarsely wavy toothed and turned under, above shiny green, hairless and rough with veins deeply sunken, beneath densely white hairy and with glands and reddish-brown scales on midvein (use hand-lens to see).

FLOWERS solitary on erect hairy stalks 1–5 in. (2.5–12.5 cm) high; white, 1–1¼ in. (2.5–3 cm) across; sepals 8–10, narrow, glandular-hairy 1/4 in. (6 mm) long; petals 8–10, white, widely spreading, 3/8–1/2 in. (10–12 mm) long.

FRUITS headlike, of many achenes 1/8 in. (3 mm) long, with persistent elongate hairy styles forming feathery plumes more than 1 in. (2.5 cm) long.

FLOWERING May–June, fruits maturing July–August.

HABITAT Arctic-alpine areas through most of Alaska except Aleutian Islands and southeastern areas.

NOTE This species is widely grown in rock gardens. Plants of this and other Alaska *Dryas* can be propagated by layering or by cuttings.

EIGHT-PETAL MOUNTAIN-AVENS

EIGHT-PETAL MOUNTAIN-AVENS

EIGHT-PETAL MOUNTAIN-AVENS

EIGHT-PETAL MOUNTAIN-AVENS

Luetkea
Luetkea pectinata (Pursh) Kuntze

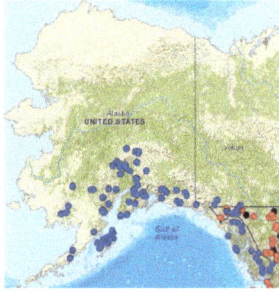

OTHER NAMES partridge-foot, meadow-spirea

DESCRIPTION Creeping and mat-forming herbaceous under-shrub, with prostrate stems and erect leafy stems 2–6 in. (5–15 cm) high.

LEAVES crowded at base, alternate above, bright green, hairless, less than 1 in. (2.5 cm) long, twice divided into 3 narrow pointed divisions.

FLOWER CLUSTERS (racemes) at top of erect leafy stems, to 2 in. (5 cm) long, bearing many small short-stalked flowers 5/16 in. (8 mm) across. **Flowers** composed of 5 pointed sepals, 5 rounded spreading white petals, about 20 stamens united at their base, and usually 5 pistils.

FRUITS 5 podlike (follicles) with several minute seeds.

FLOWERING June–September, fruit maturing July–September.

HABITAT Common, forming mats or carpets in alpine meadows near snow in mountains. Found throughout southeast and southern Alaska, west to Kodiak Island and Alaska Peninsula, north to Alaska Range.

ETYMOLOGY This genus of a single species was dedicated to Friedrich P. Lütke (1797–1882), Russian admiral and geographer who visited Alaska in 1827 on his voyage around the world.

LUETKEA

LUETKEA

Oregon Crabapple
Malus fusca (Raf.) C.K. Schneid.

OTHER NAMES western crabapple, wild crabapple

SYNONYMS *Malus diversifolia* (Bong.) M. Roemer), *Malus rivularis* (Dougl.) Roem., *Pyrus diversifolia* Bong., *Pyrus fusca* Raf.

DESCRIPTION Small deciduous tree to 25 feet (7.5 m) high, with usually several trunks to 5 in. (12.5 cm) in diameter, much branched, or a shrub forming thickets.

LEAVES with slender petioles 1–2 in. (2.5–5 cm) long, ovate, elliptic, or lance-shaped, 1½–4 in. (4–10 cm) long, short-pointed, sharply toothed and sometimes slightly 3-lobed toward tip, shiny green and becoming hairless above, beneath pale and usually slightly hairy.

TWIGS hairy when young, becoming red and shiny and later brown or gray, the side twigs or spurs short and spinelike. **Winter buds** very small, 1/16 in. (1.5 mm) long, rounded, brown, composed of many scales.

BARK gray, smooth to slightly scaly, thin.

WOOD light brown, heavy, hard, fine-textured.

FLOWER CLUSTERS (cymes) with slender stalks bearing several to many flowers 3/4 inch (2 cm) broad, composed of 5 pointed hairy sepals, 5 rounded white or pink petals, 20 stamens, and pistil with inferior 2–4-celled ovary and 2–4 styles.

FRUIT oblong, like a small apple (pome), ½–¾ in. (12–20 mm) long, yellow or red, with thin sour flesh and 2–4 papery lined cells, each with 1 or 2 large seeds.

FLOWERING in June, fruit maturing August–October.

HABITAT Commonly a thicket-forming shrub or a slow-growing small tree, scattered to plentiful on beach meadow and muskeg fringes, river bottoms, low slopes, and heavy wet soils along the Pacific coast of southeast and southern Alaska.

USES Where the trees are large enough, the wood is suitable for tool handles. It is also used for smoking salmon but less commonly than alder wood. The crabapples were eaten by Alaska Natives, and are used in jellies and preserves.

OREGON CRABAPPLE

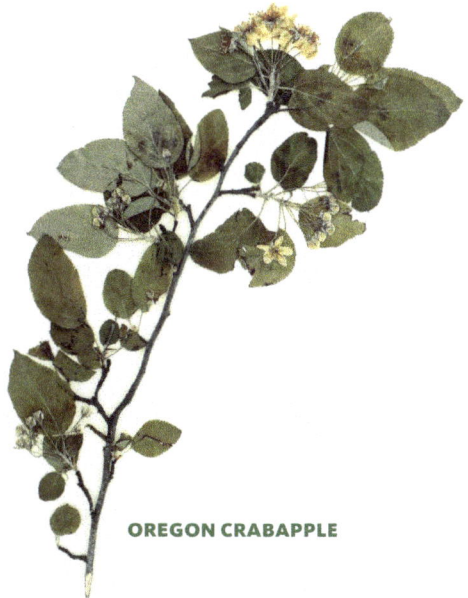

OREGON CRABAPPLE

Pacific Ninebark
Physocarpus capitatus (Pursh) Kuntze

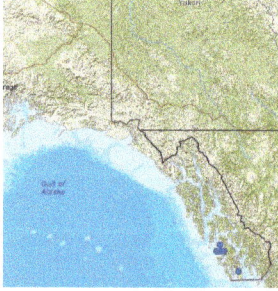

SYNONYM *Physocarpus opulifolius* (L.) Maxim. var. *tomentellus* (Ser.) Boivin.

DESCRIPTION Spreading to erect deciduous shrub 3–16 feet (1–5 m) high.

LEAVES alternate, with narrow paired stipules less than 1/4 in. (6 mm) long, shedding early, and slender petioles ½–1¼ in. (1.2–3 cm) long. **Blades** ovate to heart-shaped, 1¼–3 in. (3–7.5 cm) long and wide, palmately 3–5 lobed about half to midrib, the lobes short-pointed and irregularly or double toothed, above dark green with sparse star-shaped hairs or hairless, beneath paler and often with star-shaped hairs.

TWIGS angled, hairless or with tiny star-shaped hairs.

BARK peeling and shedding in long strips (hence the common name), exposing the orange-brown inner bark.

FLOWER CLUSTERS (corymbs) terminal, much-branched, flattened, 1½–2 in. (4–5 cm) across. **Flowers** white, nearly 1/2 in. (12 mm) across, composed of greenish cup-shaped base (hypanthium), 5 long-pointed light green persistent sepals 1/8 in. (3 mm) long, with star-shaped hairs, 5 white rounded petals about 3/16 in. (5 mm) long, about 30 stamens as long as petals or longer and 3–5 pistils slightly united at base with 1-celled ovary hairless or nearly so, 2–4 ovules, and slender style.

FRUITS 3–5 podlike (follicles), 1/4–3/8 in. (6–10 mm) long, egg-shaped, swollen, ending in long-pointed style, light brown, opening on 2 lines, persistent in winter. Seeds 2–4, more than 1/16 in. (2 mm) long, pear-shaped, shiny, light brown. Collected with flowers and fruit in July and August.

HABITAT Moist soil and streambanks near coast, uncommon and local. Extreme southeast Alaska.

NOTE Plants of related species are grown as ornamentals.

PACIFIC NINEBARK

PACIFIC NINEBARK

PACIFIC NINEBARK

CHERRY *Prunus*

Shrubs or small trees. **Leaves** alternate, simple, margins finely serrate. **Flowers** in umbel-like or raceme-like clusters; sepals 5; petals 5, white; stamens about 20; pistils 1. **Fruit** a fleshy, 1-seeded drupe. Fruits are very bitter but favored by birds, which may spread the seeds. Uncommon in Alaska.

European Bird Cherry (*Prunus padus* L.) Small introduced tree, ours likely escapes from planted trees to roadsides and disturbed areas. **Leaves** dark green, to 4 in. (10 cm) long. **Flowers** white, in clusters (racemes) that appear after the leaves emerge in spring. **Fruit** pea-sized, black, ripening in midsummer.

EURO. BIRD CHERRY

Chokecherry (*Prunus virginiana* L.) Native shrub or occasionally a small tree, often thicket-forming; **young twigs** usually hairy; **bark** smooth to fine-scaly, reddish-brown to grey-brown, the lenticels not prominent. **Leaves** finely and regularly saw-toothed, abruptly tapering to a sharp tip, green and smooth above, paler and smooth to hairy beneath; leaf stalk with 1 or 2 prominent glands near the top. **Inflorescence** a long (to 6 in [15 cm]) bottlebrush-like cluster, at the end of a short leafy spur-shoot; petals 5, nearly circular. **Fruit** 1/4–1/2 in. (6–12 mm) long, shiny, red, purple or black.

CHOKECHERRY

ROSE *Rosa*

Deciduous shrubs, sometimes climbing, with prickly or spiny twigs. **Leaves** alternate, with paired stipules attached to base of petiole, pinnate with leaflets paired except at end, toothed on edges. **Flowers** few or single, large, fragrant, composed of rounded base (hypanthium) narrowed at tip, 5 narrow sepals mostly persistent, 5 large spreading commonly pink petals broad and notched at tip, many stamens, and within the hairy base many pistils with 1-celled hairy ovary, 1 ovule, and style. **Fruit** berrylike, a rounded reddish fleshy hip containing several to many "seeds" (achenes).

ADDITIONAL SPECIES Rugosa rose (*Rosa rugosa* Thunb.), is an introduced rose reported from southern and southeast Alaska, mostly from sandy or gravelly shores. Its stems are densely covered with short, straight prickles to 3/8 in. (1 cm) long; leaves pinnate with 5–9 leaflets (most often 7), each leaflet with a distinctly corrugated (rugose) surface; flowers pink. Notable are the large, edible hips, 3/4 to 1¼ in. (2–3 cm) in diameter.

KEY TO ALASKA *ROSA*

1 Leaflets pale green and slightly hairy beneath; stipules mostly broad; prickles many; flowers 1 to few, about 2 in. (5 cm) across   **PRICKLY ROSE** (*Rosa acicularis*)
1 Leaflets simply toothed; stipules long-pointed, not toothed; twigs with slender round prickles or spines, many or scattered (interior Alaska)   **2**
2 Leaflets whitish green and mostly hairless beneath; stipules narrow; prickles few, scattered; flowers several in clusters, about 1 in. (2.5 cm) across **WOODS' ROSE** (*Rosa woodsii*)
2 Leaflets mostly doubly toothed with teeth of 2 sizes; stipules short-pointed, with gland teeth; twigs with few flattened prickles or spines paired at base of leaves or twigs (nodes); flowers mostly 1, more than 2 in. (5 cm) across (southeast and southern Alaska)
          NUTKA ROSE (*Rosa nutkana*)

Prickly Rose

Rosa acicularis Lindl.

OTHER NAME wild rose

DESCRIPTION Spiny much branched shrub to 4 feet (2.2 m) high.

LEAVES alternate, pinnate, 2–3½ in. (5–9 cm) long, with hairy glandular axis and paired broad pointed stipules 3/8–1 in. (1–2.5 cm) long. **Leaflets** mostly 5 (3–9), paired except at end, stalkless, elliptic, mostly 5/8–2¼ in. (1.5–6 cm) long and 1/4–1¼ in. (0.5–3 cm) wide, rounded at both ends, edges toothed, the teeth often gland-tipped, above dull green and usually hairless, beneath pale green and slightly hairy.

TWIGS light green when young, becoming dark red purple to gray, bristly with many straight slender gray sharp spines or prickles 1/8–1/4 in. (3–6 mm) long, unequal and round (not flattened). **Buds** 1/16–1/8 in. (2–3 mm) long, blunt, dark red, with few hairless scales.

FLOWERS 1, sometimes 2 or 3, at end of short mostly lateral twigs, on slender hairless stalk 1–1½ in. (2.5–4 cm) long, large, 1½–2¼ in. (4–6 cm) across, with 5 narrow, leaflike, greenish sepals 5/8–1¼ in. (1.5–3 cm) long, narrowest in

PRICKLY ROSE

middle, hairy and with gland hairs, and 5 pink to rose petals 3/4–1¼ in. (2–3 cm) long.

FRUIT berrylike, pearlike, elliptic or rounded, 5/8–3/4 in. (15–20 mm) long and 1/4–5/8 in. (6–15 mm) in diameter, dark red or purplish, fleshy and edible, becoming shrunken and wrinkled, curved downward and bearing at tip the persistent long sepals mostly pressed together, containing few light brown hairy "seeds" (achenes) nearly 3/16 in. (5 mm) long, persistent through winter.

FLOWERING June–July, fruits turning red in August.

HABITAT Locally common in shaded undergrowth of deciduous and spruce forests, with aspen on old burns, also thickets, roadsides, and bogs. Almost throughout central Alaska except extreme north, Aleutian and Kodiak Islands, and southeast.

USES Rose hips are eaten by grouse and other birds during fall and winter. The reddish edible fruits of this and related species, known as rose hips or rose haws, are very rich in vitamin C (ascorbic acid) and serve as a winter source. They are gathered in the fall when hard but persist through the winter, becoming soft. The juice extracted by boiling is mixed with other fruit juices or used in jellies or syrups. Jams, marmalades, and catchup are prepared from the pulp after seeds and skins are removed by sieving. Flavor is improved by combining with a tart fruit or juice such as cranberry or highbush-cranberry. It is reported that a tea has been made from the leaves.

NOTE A variable species, and hybrids with **Nootka rose** (*Rosa nutkana*) occur in southern Alaska where the ranges meet.

PRICKLY ROSE

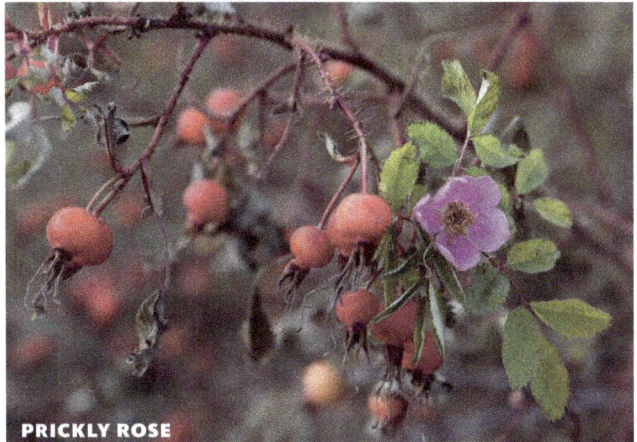
PRICKLY ROSE

Nootka Rose
Rosa nutkana K. Presl

DESCRIPTION Spiny deciduous shrub 5–8 feet (1.5–2.5 m) high, sometimes to only 2 feet (0.6 m).

LEAVES alternate, pinnate, 2½–4 in. (6–10 cm) long, with hairy glandular axis and paired short-pointed stipules 3/8–3/4 in. (1–2 cm) long with gland teeth. **Leaflets** mostly 5–7 (9), paired except at end, stalkless, elliptic or ovate, 1/2–2 in. (1.2–5 cm) long, 1/4–1½ in. (0.6–4 cm) wide, rounded at both ends, edges mostly doubly toothed with gland teeth, above dull green and hairless, beneath paler and mostly hairy along viens.

TWIGS pink brown, hairless, with few mostly paired stout flattened whitish spines 1/8–1/4 in. (3–6 mm) long, straight or slightly curved at base of leaves or twigs (nodes) or nearly spineless. **Buds** 1/8 in. (3 mm) long, blunt, dark red, with few hairless scales.

FLOWERS mostly 1, sometimes 2 or more, at end of short lateral twigs, on stout erect stalk 3/4–1 in. (2–2.5 cm) long, large, 2–2½ in. (5–6 cm) across, wth 5 narrow, leaflike, persistent sepals 5/8–1¼ in. (15–30 mm) long, narrowest in middle, hairy and with gland hairs, and 5 pink to rose petals 3/4–1¼ in. (20–30 mm) long.

FRUIT berrylike, rounded red or purplish hip 1/2–3/4 in. (12–20 mm) in diameter, without neck, with long sepals at tip, hairless, fleshy, containing several to many hairy shiny brown "seeds" (achenes) 3/16–1/4 in. (5–6 mm) long, becoming wrinkled and persistent through winter.

FLOWERING June–August, with mature fruits in August.

HABITAT Forming thickets along beaches, coastal areas of southeastern and southern Alaska, Kodiak Island, and Aleutian Islands.

NOTE Rose hips of this species are used for jelly, preserves, and as a source of vitamin C (as noted under prickly rose).

NOOTKA ROSE

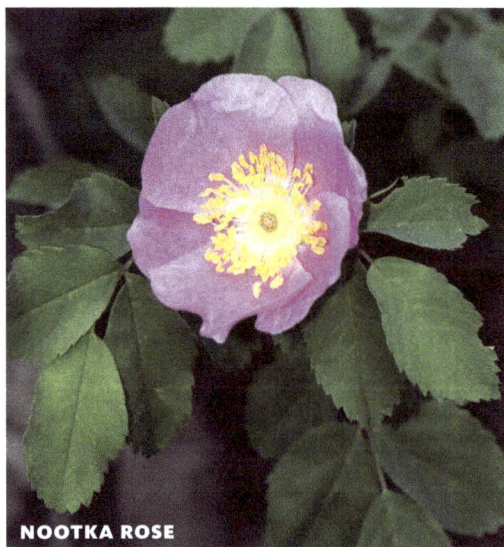

NOOTKA ROSE

Woods' Rose
Rosa woodsii Lindl.

DESCRIPTION Spiny deciduous shrub 2–5 feet (0.6–1.5 m) high. **LEAVES** alternate, pinnate, 2–4 in. (5–10 cm) long, with paired narrow pointed stipules 3/8–3/4 in. (1–2 cm) long. Leaflets 5–9, paired except at end, rounded at tip, short-pointed at base, edges toothed, above green and hairless, beneath whitish green and hairless or finely hairy.

TWIGS greenish, becoming reddish brown, hairless, with few scattered straight or curved spines or prickles 1/8–1/4 in. (3–6 mm) long.

FLOWERS mostly several in lateral clusters (cymes), sometimes few or 1, 1–1½ in. (2.5–4 cm) across, with 5 narrow, persistent sepals 3/8–3/4 in. (10–20 mm) long, mostly not glandular, and 5 light pink to rose petals 1/2–3/4 in. (12–20 mm) long.

FRUIT berrylike, rounded or elliptic hip 1/4–1/2 in. (6–12 mm) long and wide, containing many hairy "seeds" (achenes) more than 1/8 in. (3 mm) long.

FLOWERING in July.

NOTE Apparently rare and local in Alaska.

ETYMOLOGY Named for Joseph Woods (1776–1864), English botanist and specialist on roses.

WOODS' ROSE

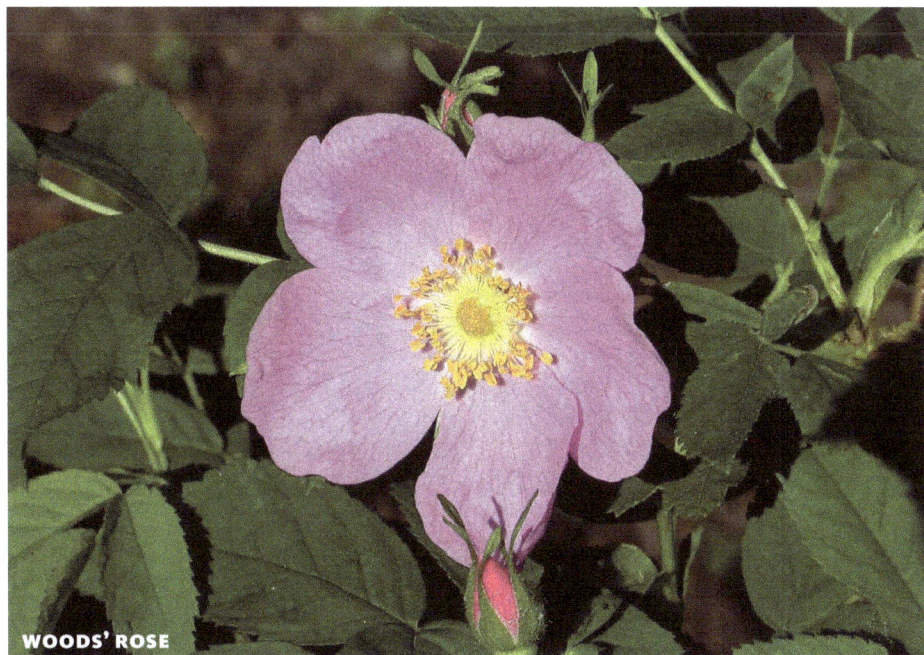

WOODS' ROSE

RASPBERRY *Rubus*

Shrubs with perennial or biennial stems, perennial herbs, and trailing vines, mostly with prickles or spines on stems and leaves. **Leaves** alternate, simple and palmately lobed or pinnately or palmately compound with 3–5 toothed leaflets, with paired stipules attached to base of petiole. **Flowers** clustered, often large, composed of saucerlike to conic base (hypanthium), calyx of 5 persistent sepals, 5 white to red petals, many stamens, and many pistils with 1-celled ovary, 2 ovules, and style. **Fruits** aggregate, composed of usually many separate drupelets, fleshy, mostly edible, 1-seeded. Major Alaskan species are four *Rubus* with woody stems (described below); an additional three non-woody species less than 1 feet (0.3 m) high, are briefly described.

KEY TO SHRUBBY ALASKA *RUBUS*

1   Leaves simple, palmately 3–7 lobed; stems erect, without spines or prickles; fruit red, half round, edible   **WESTERN THIMBLEBERRY** (*Rubus parviflorus*)

1   Leaves compound, with 3 or 5 leaflets; stems often spreading, spiny or prickly; fruit rounded   **2**

2   Twigs covered with bristles and prickles; fruit a red raspberry   **AMERICAN RED RASPBERRY** (*Rubus idaeus*)

2   Twigs with spines or prickles   **3**

3   Twigs light brown, zigzag, with weak straight rounded prickles; petals pink to purple; fruit yellow to dark red, edible, almost tasteless   **SALMONBERRY** (*Rubus spectabilis*)

3   Twigs whitish, with stout hooked flattened prickles or spines; petals white; fruit reddish to black raspberry   **WESTERN BLACK RASPBERRY** (*Rubus leucodermis*)

Common Red Raspberry

Rubus idaeus L.

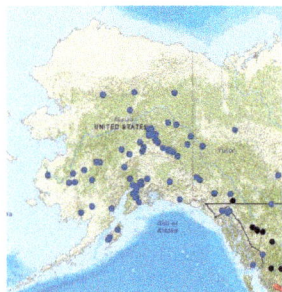

OTHER NAMES red raspberry, raspberry

SYNONYMS *Rubus strigosus* Michx.; Alaska plants are subsp. *strigosus* (Michx.) Focke

DESCRIPTION Deciduous bristly shrub 2–4 feet (0.6–1.2 m) high with biennial stems.

LEAVES pinnately compound, 2½–7 in. (6–18 cm) long, with very narrow paired stipules less than 3/8 in. (1 cm) long. **Leaflets** 3 or 5, paired except at end, ovate, 1½–3½ in. (4–9 cm) long, 3/4–2 in. (2–5 cm) wide, long-pointed at tip, rounded at base, irregularly toothed and shallowly lobed, above green and mostly hairless, beneath gray green and usually hairy.

TWIGS reddish brown, covered with bristles and prickles, often hairy.

BARK yellow brown, shreddy.

FLOWERS 1–4 lateral, small, 3/8–1/2 in. (10–12 mm) across, composed of calyx of 5 narrow hairy sepals about 1/4 in. (6 mm) long, 5 white oblong petals about 1/4 in. (6 mm) long erect or slightly spreading, many (75–100) stamens, and many pistils.

FRUIT aggregate, a red raspberry, rounded, 3/4 in. (2 cm) long and broad, of many hairy drupelets, separating from its base.

COMMON RED RASPBERRY

FLOWERING June–July, fruits maturing July–September.
HABITAT Common to abundant in openings and borders of forests, forming thickets, also a roadside weed. Across Alaska from interior to southeast (but not in far north).
NOTE Red raspberries are eaten fresh or in jams and jellies.

COMMON RED RASPBERRY

Western Black Raspberry
Rubus leucodermis Dougl. ex Torr. & Gray

OTHER NAME whitebark raspberry
DESCRIPTION Deciduous spiny shrub 3–6 feet (1–2 m) high, with biennial stems.
LEAVES compound, 3–5 in. (7.5–12.5 cm) long, with very narrow paired stipules less than 1/4 in. (6 mm) long. **Leaflets** 3, ovate, 3/4–3 in. (2–7.5 cm) long, 3/8–2 in. (1–5 cm) wide, short to long-pointed at tip, rounded at base, edges irregularly toothed to shallowly lobed, above green and hairless or nearly so, beneath whitish hairy.
TWIGS whitish, with many stout hooked flattened prickles or spines to 1/4 in. (6 mm) long.
FLOWER CLUSTERS (racemes) of 2–7 flowers close together at leaf base, less than 1/2 in. (12 mm) across, composed of calyx of 5 narrow hairy sepals 1/4–1/2 in. (6–12 mm) long and bent downward, 5 white petals shorter than sepals, many (70–100) stamens, and many pistils.
FRUIT aggregate, a reddish to black raspberry with whitish bloom, rounded, to 1/2 in. (12 mm) broad, of many hairy drupes, edible, separating from base. Collected with fruit in August and September.
HABITAT Reported from southeast Alaska.

WESTERN BLACK RASPBERRY

WESTERN BLACK RASPBERRY

Western Thimbleberry
Rubus parviflorus Nutt.

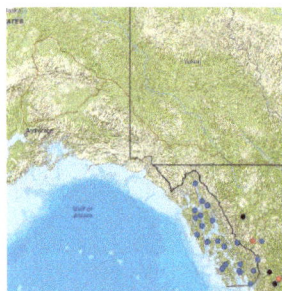

OTHER NAMES thimbleberry

SYNONYM *Rubus nutkanus* Moc.

DESCRIPTION Deciduous erect shrub 2–5 feet (0.6–1.5 m) high, with erect perennial stems, without spines.

LEAVES simple, with paired lance-shaped stipules 1/4–1/2 in. (6–12 mm) long and slender petioles 1–4 in. (2.5–10 cm) long with stalked gland hairs. **Blades** rounded or 5-angled, 2½ –6 in. (6–15 cm) long and broad, thin, palmately lobed with mostly 5 (sometimes 3 or 7) shallow short-pointed lobes, heart-shaped at base, edges sharply doubly toothed with gland teeth, with 5 main veins from base, above dull green and nearly hairless, beneath paler, slightly hairy and with stalked gland hairs along veins.

TWIGS greenish, finely hairy and with stalked gland hairs.

BARK gray, shreddy or flaky.

FLOWER CLUSTERS (corymbs or panicles) terminal and flat-topped. **Flowers** mostly 3–7, 1½ –2 in. (4–5 cm) across; composed of calyx of 5 spreading, narrow, hairy, greenish sepals 3/8–5/8 in.

WESTERN THIMBLEBERRY

(10–15 mm) long; 5 white, obovate, spreading petals 5/8–1 in. (20–25 mm) long; many stamens, and many pistils.

FRUIT aggregate, thimblelike, half round and flattened, 1/2 in. (12 mm) across, juicy and edible, composed of many small hairy red drupelets 1/16 in. (2 mm) long.

FLOWERING June–July, with mature fruits August–September.

HABITAT Common in moist soil in thickets and openings of forests, along roadsides, and on cut-over land, southeast Alaska north to Lynn Canal and Yakutat. Southeast Alaska east to Ontario and Michigan, south in mountains to New Mexico, California, and northern Mexico.

USES The fruits are excellent for jelly but too seedy for jam.

WESTERN THIMBLEBERRY

WESTERN THIMBLEBERRY - FRUIT

Salmonberry
Rubus spectabilis Pursh

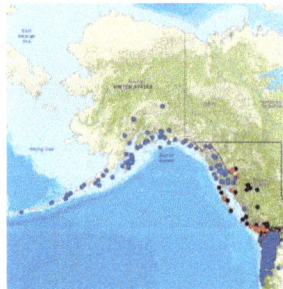

DESCRIPTION Large or small, thicket-forming deciduous shrub 2–7 feet (0.6–2 m) high, with erect, curved, biennial stems. **LEAVES** compound, 2–5 in. (5–12 cm) long, slender hairy axis and paired needlelike, hairy, persistent stipules 1/4–3/8 in. (6–10 mm) long. **Leaflets** 3, ovate, mostly 1–2½ in. (2.5–6 cm) long and 5/8–2 in. (1.5–2.5 cm) wide, the terminal one larger than the lateral pair, thin, long-pointed at tip and short-pointed at base, sharply and irregularly toothed and shallowly lobed, above green and nearly hairless, beneath paler and slightly hairy.

TWIGS zigzag, light brown, becoming hairless, often with scattered sharp weak spines or prickles 1/16–1/8 in. (2–3 mm) which break off easily.

BARK light brown, becoming shreddy.

FLOWERS 1 or 2 lateral on long slender stalks, large and showy, 1½ in. (4 cm) across, composed of calyx of 5 spreading long-pointed hairy greenish sepals 3/8–5/8 in. (10–15 mm) long, 5 spreading elliptic pink to reddish purple petals 5/8–7/8 in. (15–22 mm) long, many (75–100) purplish stamens, and many (20–40) pistils.

FRUIT aggregate, separating from base and persistent calyx like a raspberry, orange to dark red, conelike, 5/8–1 in. (1.5–2.5 cm) long and broad, juicy, of many small drupes, edible, mild-tasting.

FLOWERING April–July, maturing fruit by early July near Ketchikan, but not until August on Kodiak and Afognak Islands and at higher altitudes.

HABITAT Salmonberry is scattered to common or abundant in moist soil, forming dense

SALMONBERRY

SALMONBERRY

SALMONBERRY

thickets in openings in lowland forests, clearings, and along streams. It spreads quickly after clearcutting and can be a serious competitor of conifer regeneration on moist, valley-bottom sites. Found in southeast and southern Alaska, west to Aleutian Islands.

USES The fruits make good jelly but are rather seedy for jam. They are eaten by bears in the fall. New leaves and twigs are browsed in the spring by deer, moose, and mountain goats.

ADDITIONAL SPECIES Besides the four shrubby species with woody stems described here, three additional native species of this genus are herbs with creeping stems or erect herbaceous stems usually less than 1 feet (30 cm) high:

Nagoon-berry (*Rubus arcticus* L.; other names: wineberry, Arctic bramble, kneshenada; synonym *Rubus acaulis* Michx.). Herbs 2–10 in. (5–25 cm) high from spreading rootstock. Leaves with slender petioles 1–2 in. (2.5–5 cm) long and 3 almost stalkless elliptic toothed leaflets 5/8–1¼ in. (1.5–3 cm) long; a variation with simple rounded leaves 1–2 in. (2.5–5 cm) long and broad, deeply 3-lobed. Flowers 1–3, pink or red, 3/4–1¼ in. (2–3 cm) across. Fruit red, 1/2–3/4 in. (1.2–2 cm) across, of 15–40 drupelets not separating from calyx, edible. Common in sedge meadows and bogs, interior, western, southern, and southeast Alaska and through Aleutian Islands. The berries are a favorite for jam, jelly, and wine, because of their excellent flavor. This widespread, variable species has intergrades and hybrids among its races.

Cloudberry (*Rubus chamaemorus* L., other name baked-apple berry) Erect herb 2–8 in. (5–20 cm) high from creeping rootstock. Leaves 2 or 3 with slender petioles 1/2–1 in. (1.2–2.5 cm) long and rounded blades 1–2 in. (2.5–5 cm) across, with 3 or 5 rounded lobes and finely toothed border. Flower 1, erect, white, 1/2–1 in. (1.2–2.5 cm) across. Fruit 1/2–3/4 in. (1.2–2 cm) in diameter, edible, composed of 6–18 large pink drupelets the color of a baked apple. In bogs almost throughout Alaska, across Canada to Labrador and Greenland, south to New York. Also across northern Eurasia. The edible berries are collected in quantities in late August and early September and are stored frozen by the Eskimos for winter use. The berries are eaten fresh and in jam, shortcake, and pie. Rich in vitamin C, even when frozen soon after picking.

Five-leaf bramble (*Rubus pedatus* Sm.). Slender trailing herbaceous vine rooting at nodes and forming mats, flowering twigs less than 1 in. (2.5 cm) high. Leaves 2–4, palmately compound, with slender petiole 1–3 in. (2.5–7.5 cm) long and 5 nearly stalkless obovate irregularly toothed leaflets 3/8–1¼ in. (1–3 cm) long. Flower 1, erect white, 1/2–5/8 in. (1.2–1.5 cm) across, with petals and sepals about equal. Fruit of 1–6 red drupelets 3/8 in. long (1 cm.), juicy, edible, used for jam. Forests in southern and southeast Alaska, southeast to Alberta, Montana, and Oregon; also in Japan.

  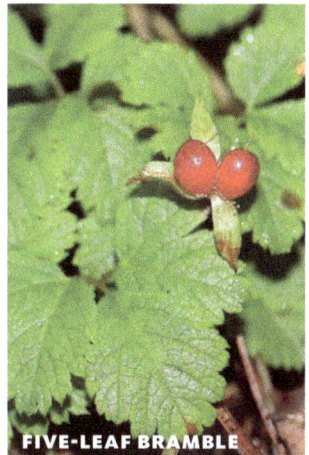

NAGOON-BERRY   CLOUDBERRY   FIVE-LEAF BRAMBLE

MOUNTAIN-ASH *Sorbus*

Deciduous shrubs and small trees with stout twigs and large buds with overlapping scales. **Leaves** alternate, with paired stipules attached to petiole, pinnate with 7–17 toothed leaflets paired except at end. **Flower clusters** (corymbs) terminal, much branched, showy. **Flowers** many, small, white, composed of calyx of 5 triangular persistent sepals, 5 white mostly rounded petals, 15–20 stamens, and pistil with inferior 2–5-celled ovary, 2 ovules in each cell, and 2–5 styles. **Fruits** like a small red apple (pome) with calyx at tip, 2–4-celled with 1–2 flattened seeds in each cell. Alaska has three native species, two of which become small trees, and one introduced and naturalized tree species.

KEY TO ALASKA MOUNTAIN-ASH

1 Leaflets mostly 7 or 9, lance-shaped, long-pointed; westernmost Aleutian Islands only
SIBERIAN MOUNTAIN-ASH (*Sorbus sambucifolia*)
1 Leaflets 9 or 11 or more, oblong or elliptic, short-pointed or rounded at tip **2**
2 Leaflets mostly 9 or 11, elliptic, rounded or short-pointed at tip, edges not toothed in lowest third; southern and southwestern Alaska
SITKA MOUNTAIN-ASH (*Sorbus sitchensis*)
2 Leaflets mostly 11–15, oblong, short-pointed, edges toothed nearly to base **3**
3 Leaflets becoming hairless; shrub or rarely small tree
CASCADE MOUNTAIN-ASH (*Sorbus scopulina*)
3 Leaflets white-hairy beneath; naturalized tree
EUROPEAN MOUNTAIN-ASH (*Sorbus aucuparia*)

European Mountain-Ash

Sorbus aucuparia L.

OTHER NAME Rowan-tree

DESCRIPTION Deciduous small to medium tree planted as an ornamental in southeast Alaska, introduced and sparingly naturalized, 20–40 feet (6–12 m) tall and 1 feet (30 cm) in trunk diameter, with symmetrical rounded crown.

LEAVES pinnate, 4–8 in. (10–20 cm) long, with paired 3-angled stipules. **Leaflets** 9–17, oblong or lance-shaped, 1–2 in. (2.5–5 cm) long, short-pointed, with edges coarsely toothed except near unequal rounded base, dull green and

becoming hairless above, pale and white-hairy beneath.

YOUNG TWIGS and winter buds densely white-hairy or woolly, winter buds conical, 3/16–3/8 in. (5–10 mm) long.

BARK dark gray, smooth, with horizontal lines (lenticels), aromatic.

FLOWER CLUSTERS (corymbs) terminal rounded, 4–6 in. (10–15 cm) across, bearing 75–100 flowers on densely white-hairy stalks. **Flowers** 3/8 in. (10 mm)

EUROPEAN
MOUNTAIN-ASH

across, composed of 5 triangular white-hairy sepals, 5 white rounded petals 3/16 in. (4 mm) long, many stamens, and pistil with inferior hairy ovary and 3–4 styles.

FRUITS many, like a small apple (pome), **seeds** elliptic, light brown, 3/16 in. (4 mm) long. Fruits maturing in August–September.

HABITAT Planted as an ornamental tree at Wrangell, Ketchikan, Sitka, Juneau, and other towns along the coast of southeast Alaska, where it spreads from cultivation. Sparingly naturalized along roads and forming thickets. Widely planted and naturalized in many places across Canada and northern United States.

NOTE Not a true ash, European mountain-ash is the only introduced or exotic tree to become established in Alaska and grow as if wild. The fruits persist into late fall and early winter and provide food for birds, such as crossbills, grosbeaks, and cedar waxwings, which probably spread the seeds; crows also eat the fruits. Its specific name, meaning to catch birds, refers to the use of the mucilaginous fruits by fowlers in making birdlime (an adhesive substance used to trap birds).

EUROPEAN MOUNTAIN-ASH

EUROPEAN MOUNTAIN-ASH

Siberian Mountain-Ash
Sorbus sambucifolia (Cham. & Schlecht.) M. Roemer

OTHER NAMES elder-leaf mountain-ash

SYNONYM *Pyrus sambucifolia* Cham. & Schlecht.

DESCRIPTION Deciduous shrub 2–5 feet (0.6–1.5 m) high.

LEAVES pinnate, 2½–5 in. (6–12.5 cm) long, with paired rusty hairy lance-shaped stipules 1/8 in. (3 mm) long. **Leaflets** 7 or 9 (11), lance-shaped, 1–1¾ in. (2.5–4.5 cm) long and 3/8–3/4 in. (1–2 cm) wide, usually broadest near unequal rounded base, gradually narrowed to long-pointed tip, edges sharply toothed almost to base, becoming hairless, above shiny green, beneath dull and paler.

TWIGS rusty hairy when young, becoming gray, with few elliptic whitish dots (lenticels). **Buds** shiny reddish brown, sticky, slightly rusty hairy.

FLOWER CLUSTERS (corymbs) terminal, rounded, 1¼–2 in. (3–5 cm) wide, bearing 8–15 flowers on slightly rusty hairy stalks. **Flowers** 3/8–5/8 in. (1–1.5 cm) across, composed of 5 triangular sepals hairy on edges, 5 white rounded petals 3/16 in. (5 mm) long, many stamens, and pistil with inferior hairy 5-celled ovary and 5 styles.

FRUITS few, like a small apple (pome), elliptic, 3/8–5/8 in. (10–15 mm) in diameter, reddish with a bloom, with calyx at tip, containing few dark brown seeds more than 1/8 in. (3 mm) long.

FLOWERING in July, fruits maturing in August.

HABITAT In Alaska found only on westernmost Aleutian Islands (Attu, Buldir, Alaid, Agattu); main range Asia from Kamchatka to Korea and Japan.

NOTE The fruits are described as not very acid and suitable for jam.

SIBERIAN MOUNTAIN-ASH

SIBERIAN MOUNTAIN-ASH

Cascade Mountain-Ash
Sorbus scopulina Greene

OTHER NAMES Greene's mountain-ash, western mountain-ash

SYNONYMS *Sorbus alaskana* G.N. Jones not Hollick, *Pyrus scopulina* (Greene) Longyear

DESCRIPTION Deciduous shrub 3–13 feet (1–4 m) high, rarely becoming a small tree to 20 feet (6 m) high and 4 in. (10 cm) d.b.h.

LEAVES pinnate, 4–9 in. (10–23 cm) long, with paired, very narrow hairless stipules 1/4–3/8 in. (6–10 mm) long. **Leaflets** 11–15, stalkless, oblong-lanceolate, 1¼–2½ in. (3–6 cm) long and 3/8–3/4 in. (1–2 cm) wide, unequal and rounded at base, short or long-pointed at tip, edges sharply toothed almost to base, becoming hairless, above shiny dark green, beneath slightly paler.

TWIGS with whitish hairs when young, with scattered elliptic dots (lenticels). **Buds** conical, dark reddish brown, inner scales with whitish hairs.

BARK gray, smooth.

FLOWER CLUSTERS (corymbs) terminal, rounded, 1¼–3 in. (3–7.5 cm) broad, bearing on whitish hairy stalks many fragrant flowers 3/8 in. (1 cm) across, composed of 5 minute triangular sepals, 5 elliptic petals 3/16 in. (5 mm) long, many stamens; and pistil with inferior, hairy, 3–4-celled ovary and 3–4 styles.

FRUITS fewer than 25, like a small apple (pome), round, less than 3/8 in. (10 mm) in diameter, bright shiny red, bitter, with few elliptic brown seeds more than 1/8 in. (3 mm) long, persistent in winter.

FLOWERING June–July, maturing fruits in July.

HABITAT Openings and clearings in forests.

NOTE This normally shrubby species is found at Haines as a small tree 20 feet (6 m) high.

CASCADE MOUNTAIN-ASH

CASCADE MOUNTAIN-ASH

Sitka Mountain-Ash
Sorbus sitchensis M. Roemer

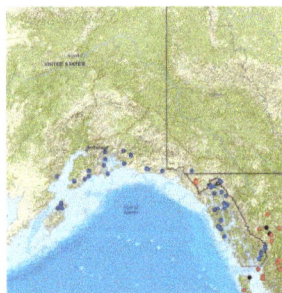

OTHER NAMES western mountain-ash, Pacific mountain-ash
DESCRIPTION Deciduous shrub 4–8 feet (1.2–2.5 m) high, or a small tree to 15–20 feet (4.5–6 m) high and 6 in. (15 cm) in trunk diameter, with handsome, round-topped head. In rocky alpine situations at higher altitudes it is a low shrub often only 1–2 feet (30–61 cm) high.
LEAVES pinnate, 4–8 in. (10–20 cm) long, with paired narrow rusty-hairy stipules. **Leaflets** usually 9 or 11 (sometimes 7 to 13), elliptic or oblong, 1¼ –2½ in. (3–6 cm) long, rounded or blunt-pointed at tip, with edges coarsely and sharply toothed above the middle, dull blue green and hairless above, pale and hairless or nearly so beneath.
TWIGS stout, rusty hairy when young, becoming gray, with few elliptic dots (lenticels), with odor and bitter taste of cherry. **Buds** oblong, to 1/2 in. (12 mm) long, dull reddish brown, densely rusty hairy.
BARK gray, smooth.
WOOD pale brown, light-weight, fine-textured.
FLOWER CLUSTERS (corymbs) terminal, rounded, 2–4 in. (5–10 cm) across, bearing 15–60 flowers on rusty-hairy stalks. **Flowers** small, 1/4 in. (6 mm) across, fragrant, composed of 5 broadly triangular hairless sepals, 5 white rounded petals 1/4 in. (5 mm) long, many stamens, and pistil with inferior hairy ovary and 3–4 styles.
FRUITS several to many, like a small apple (pome), round, 3/8–1/2 in. (10–12 mm) in diameter, red but becoming orange and purple, with few elliptic brown seeds 1/8 in. (3 mm) long.
FLOWERING June–August, fruits maturing in August–September.
HABITAT Uncommon to rare in forests from sea level to timberline, Pacific coast of southeast and southern Alaska.
USES Often cultivated as an ornamental north to Anchorage but with less regular form than European mountain-ash. Birds eat the fruits.
NOTE Hybrids with Greene mountain-ash (*Sorbus scopulina*) have been reported.
ETYMOLOGY Sitka mountain-ash is named for Sitka, Alaska, where it was discovered.

SITKA MOUNTAIN-ASH

SITKA MOUNTAIN-ASH

SITKA MOUNTAIN-ASH

SPIRAEA *Spiraea*

Deciduous shrubs with alternate simple small leaves, short petioles, and blades with toothed edges, without stipules. **Flowers** many in much-branched terminal clusters, small, with cup-shaped base (hypanthium), 5 persistent sepals, 5 rounded white or pink petals, many stamens, and mostly 5 distinct pistils composed of 1-celled ovary, 2–several ovules, and slender persistent style. **Fruits** mostly 5 podlike (follicles), splitting open on 1 line, containing 2–several minute seeds.

KEY TO ALASKA SPIRAEA

1   Flower clusters flat-topped to half round, petals white; leaves rounded at both ends, with edges mostly toothed nearly to base   **BEAUVERD SPIRAEA** (*Spiraea stevenii*)
1   Flower clusters conic, much longer than broad, petals pink to rose; leaves short-pointed to rounded at both ends, with edges toothed in upper half
                                        DOUGLAS SPIRAEA (*Spiraea douglasii*)

Douglas Spiraea
Spiraea douglasii Hook.

OTHER NAMES Menzies' spiraea

SYNONYMS *Spiraea menziesii* Hook.

DESCRIPTION Erect deciduous shrub 3–5 feet (1–1.5 m) high. **LEAVES** with short hairy petioles about 1/8 in. (3 mm) long. Blades elliptic to oblong, 1¼ –3 in. (3–7.5 cm) long and 3/8– 1¼ in. (1–3 cm) wide, short-pointed to rounded at both ends, edges sharply toothed in upper half, above dark green and usually hairless, beneath pale green and sometimes hairy. **TWIGS** slender, reddish brown, with fine soft hairs when young, sometimes nearly hairless, becoming dark brown and hairless. **Buds** 1/16 in. (2 mm) long, scaly, white hairy toward tip.

FLOWER CLUSTERS (panicles) terminal, 1½ –6 in. (4–15 cm) long, conic, several times as long as broad, mostly finely hairy. **Flowers** many, crowded, short-stalked, small, 1/4 in. (6 mm) across, with 5 triangular sepals bent down, 5 pink to rose petals, round to obovate, 1/16 in. (2 mm) long, many pink to rose stamens, and 5 pistils.

FRUITS 5 podlike (follicles) 1/8 in. (3 mm) long, shiny brown, hairless or nearly so, containing 2 to several narrow seeds, persistent in winter. Collected in flower in July and August, with mature fruits in September.

HABITAT Moist soil, especially borders of streams and lakes. Ketchikan and elsewhere in extreme southeastern Alaska.

ETYMOLOGY Named for the discoverer, David Douglas (1798–1834), Scotch botanical explorer. Plants of Alaska and adjacent coast of British Columbia were formerly accepted as a separate species, Menzies' spirea (*Spiraea menziesii* Hook.), later reduced to a variety *Spiraea douglasii* var. *menziesii*).

DOUGLAS SPIRAEA

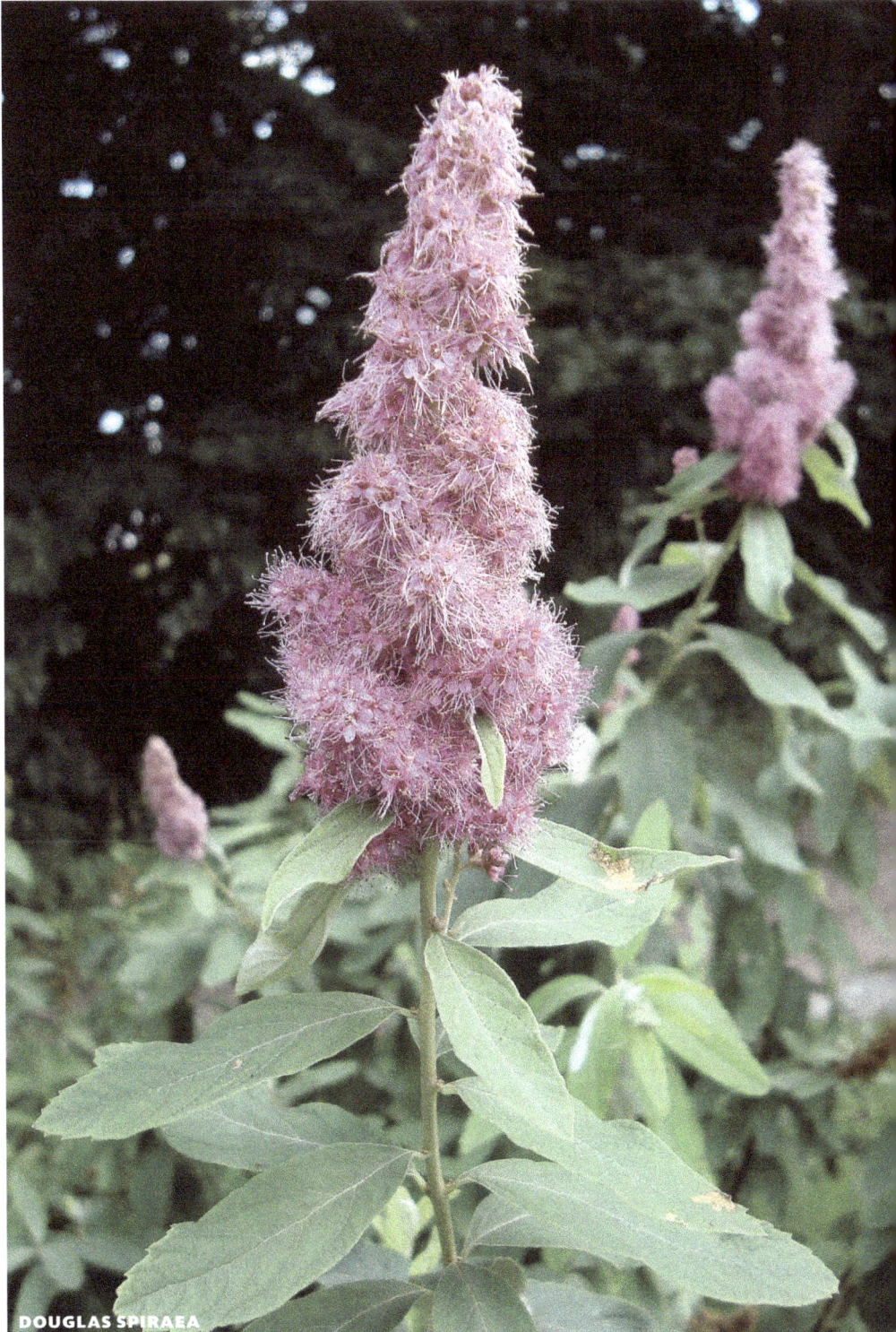

DOUGLAS SPIRAEA

Beauverd Spiraea
Spiraea stevenii (C.K. Schneid.) Rydb.

OTHER NAME Alaska spiraea

SYNONYM *Spiraea beauverdiana* Schneid.

DESCRIPTION Small much-branched deciduous shrub 1–2 (4) feet (3–6 (12) dm) high.

LEAVES with short petioles 1/16 in. (2 mm) long. **Blades** elliptic to ovate, 5/8–2 in. (1.5–5 cm) long, 3/8–1¼ in. (1–3 cm) wide, rounded at both ends, edges sharply toothed nearly to base (sometimes almost without teeth), above dull green and hairless or nearly so, beneath paler and often finely hairy.

TWIGS slender, purplish brown, hairy when young, afterwards outer bark shedding in long thin strips. **Buds** about 1/16 in. (2 mm) long, of few slightly hairy scales.

FLOWER CLUSTERS (corymbs or headlike) terminal, flattened to half round, 3/4–1½ in. (2–4 cm) across. **Flowers** many, crowded, shortstalked, small, about 1/4 in. (6 mm) across, with 5 triangular sepals bent down, 5 white petals (or rose-tinged in center, pink in bud) 1/16 in. (2 mm) long, many white stamens more than twice as long as petals, and 5 pistils. **FRUITS** usually 5 podlike (follicles) less than 1/8 in. (3 mm) long, shiny brown, finely hairy, containing 2–several narrow seeds, persistent in winter.

FLOWERING June–August, with mature fruits July–September.

HABITAT Common in tundra, swamps, black spruce muskegs, and forests, from lowland to alpine. Almost throughout Alaska except extreme north, Aleutian Islands, and southeast region.

NOTE A variable species; plants at high altitudes are dwarfed (often less than 8 in. (2 dm) high) with small leaves.

ETYMOLOGY This species honors Gustave Beauverd (1867–1942), a Swiss botanist.

BEAUVERD SPIRAEA

BEAUVERD SPIRAEA

BEAUVERD SPIRAEA

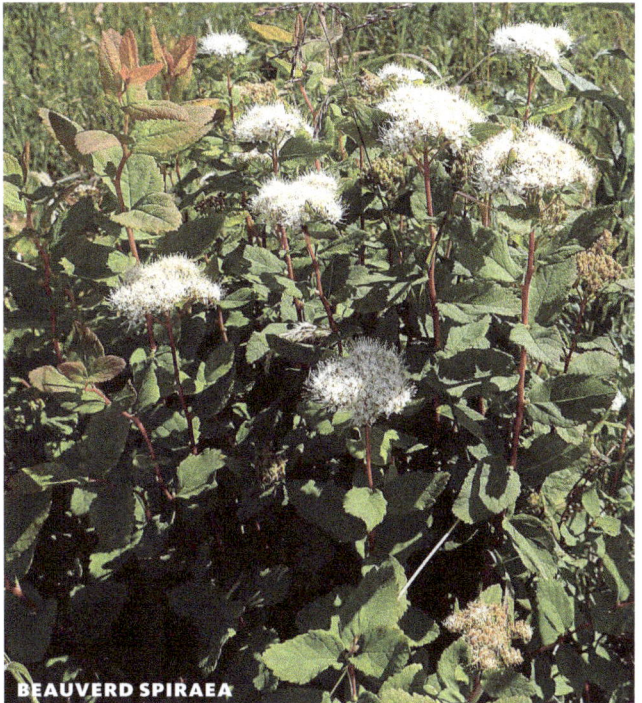

BEAUVERD SPIRAEA

■ WILLOW FAMILY *Salicaceae*

The willow family (Salicaceae) contains the cottonwoods, poplars, and aspens (the genus *Populus* with three tree species in Alaska), and the willows (*Salix*), a large genus of 30 (or more) native species ranging in size from creeping or dwarf shrubs to large shrubs and small trees. Distinguishing characters are:
• **leaves** borne singly (alternate), with margins evenly toothed or without teeth (entire) but not lobed;
• **flower clusters** (catkins) composed of an axis bearing many small flowers each above a scale, in early spring before or with the leaves;
• **flowers** without sepals or petals, of 2 kinds on different plants, male flowers with pollen and on other plants the female flowers with seeds; and
• the tiny **seeds** with long white cottony hairs, borne in small seed capsules mostly 2-parted.

Cottonwoods, poplars, and aspens usually have broad leaves with petiole nearly as long as the blade, stout twigs, and large winter buds with several scales exposed, resinous (except in aspen), and a terminal bud present. Willows usually have narrow leaves with very short petioles, slender or wiry twigs, and small winter buds covered by a single scale, the terminal bud absent. Catkins in *Populus* hang down, while those of willows are upright or slightly spreading. Flowers of cottonwoods have deeply lobed scales soon shedding, a broad or cup-shaped disk, and 10 to many stamens. Willow flowers have scales without or with teeth, are persistent or late shedding, the disk reduced usually to one small gland, and 2–8 stamens.

COTTONWOOD, POPLAR, ASPEN *Populus*

This genus has no single English common name. The three Alaska species, all common trees, are balsam poplar, black cottonwood, and quaking aspen.

KEY TO ALASKA *POPULUS*

1 Leaf blades nearly round, less than 2 in. (5 cm) long; petioles flattened
QUAKING ASPEN (*Populus tremuloides*)
1 Leaf blades longer than broad, 2½ –5 in. (6–12.5 cm) long; petioles round in section   2
2 Seed capsules pointed, hairless, 2-parted; leaves pale green and brownish beneath; tree of interior forests   BALSAM POPLAR (*Populus balsamifera*)
2 Seed capsules rounded, hairy, 3-parted; leaves whitish beneath; tree of coastal forests
BLACK COTTONWOOD (*Populus trichocarpa*)

Balsam Poplar
Populus balsamifera L.

OTHER NAMES tacamahac, tacamahac poplar, cottonwood
SYNONYM *Populus tacamahaca* Mill.
DESCRIPTION Medium-sized deciduous tree usually 30–50 feet (9–15 m) high, with straight trunk 4–12 in. (10–30 cm) in diameter and long thin open crown, sometimes a large tree 80–100 feet (24–30 m) tall and 2 feet (60 cm) in trunk diameter.
LEAVES with slender petioles 1–2 in. (2.5–5 cm) long, round, finely hairy. **Leaf blades** ovate or broadly lance-shaped, 2½ – 4½ in. (6–11 cm) long, 1½ –3 in. (4–7.5 cm) wide, mostly long-pointed at tip and rounded at base, with many small rounded teeth, hairless or nearly so, shiny dark green above, pale green and rusty brown beneath.

TWIGS red brown and hairy when young, with orange dots (lenticels), becoming gray, with raised leaf scars showing 3 dots. **Winter buds** large, to 1 in. (2.5 cm) long, long-pointed, sticky or resinous, covered with shiny brown scales, with pungent balsam odor which permeates the air in spring.

BARK light gray to gray, smooth, becoming rough, thick, and deeply furrowed.

WOOD with thick whitish sapwood and light brown heartwood, fine-textured, lightweight, soft.

FLOWER CLUSTERS (catkins) 2–3½ in. (5–9 cm) long, narrow, drooping, with many small flowers about 1/8 in. (3 mm) long, each with disk and above a light brown hairy lobed scale, male and female on different trees (dioecious). **Male flowers** with 20–30 reddish purple stamens; **female flowers** with conic slightly 2-lobed hairless ovary and 2 broad wavy stigmas.

SEED CAPSULES in catkins to 6 in. (15 cm) long, short-stalked, egg-shaped, 1/4–5/16 in. (6–8 mm) long, long-pointed, light brown, hairless but warty, 2-parted, with many tiny cottony seeds.

FLOWERING in May–June before the leaves, fruit maturing in June.

HABITAT Common in river valleys including sandy bottoms and gravelly floodplains, terraces, and coarse alluvial fans throughout the interior except near the coasts. In forests, especially in openings and clearings, it is associated with white spruce, birch, and aspen. It is often common with willows and alders in floodplain thickets and along river banks.

In the mountains balsam poplar extends to somewhat higher altitudes than white spruce, to 3,500 feet altitude or more on north and west slopes of the Alaska Range. Also, it ranges farther north to the Arctic slope in a few places. At Firth River near the northeast corner of Alaska and north of the tree-line, balsam poplar, white spruce, and feltleaf willow are the only tree species.

USES Balsam poplar, sometimes erroneously called balm-of-Gilead, is a rapidly growing tree. The wood is used chiefly for boxes, crates, and pulpwood southward. A small amount is sawn for use in the Anchorage area.

NOTE Balsam poplar intergrades or hybridizes with black poplar in southern Alaska where ranges of the two overlap, as mentioned under the latter. Rare hybrids with quaking aspen, which has smaller, rounded leaves, and flattened petioles, have also been recorded.

BALSAM POPLAR

BALSAM POPLAR

Quaking Aspen
Populus tremuloides Michx.

OTHER NAMES American aspen, trembling aspen, popple

DESCRIPTION Small to medium-sized deciduous tree commonly 20–40 feet (6–12 m) tall, maximum 80 feet (24 m), with straight trunk 3–12 in. (7.5–30 cm) in diameter, maximum 18 in. (46 cm), and short, irregularly bent limbs making a narrow domelike crown.

LEAVES with slender flattened petioles 1–2.5 in. (2.5–6 cm) long. **Leaf blades** nearly round, 1–2 in. (2.5–5 cm) long and broad, short-pointed at tip, rounded at base, with many small rounded teeth, hairless, shiny green above, pale beneath, which tremble in the slightest breeze, turning bright yellow (sometimes reddish) in autumn.

TWIGS slender, reddish and slightly hairy when young, becoming gray, with raised leaf scars showing 3 dots. **Winter buds** conic, 1/4 in. (6 mm) long, long-pointed, of shiny red brown hairless scales, not resinous or flower buds slightly so.

BARK whitish or greenish gray, smooth, thin, with characteristic curved scars and black knots.

QUAKING ASPEN

WOOD of broad whitish sapwood and light brown heartwood, fine-textured, lightweight, soft, and brittle.

FLOWER CLUSTERS (catkins) 1–2½ in. (2.5–6 cm) long, narrow, drooping, with many small flowers 1/8 in. (3 mm) long, each with saucer-shaped disk and above a brown hairy lobed scale, male and female on different trees (dioecious). **Male flowers** with 6–12 stamens; **female flowers** with conic ovary, short style, and 2 stigmas each 2-lobed.

SEED CAPSULES in catkins 3–4 1/2 in. (7.5–11 cm) long, nearly stalkless, less than 1/4 in. (6 mm) long, conic, hairless, 2-parted, with many tiny cottony seeds.

FLOWERING in May before the leaves, fruit maturing in May–June.

HABITAT Common on south slopes, well-drained benches, and creek bottoms throughout the interior of Alaska to about 3,000 feet altitude. It often occurs in dense pure stands, especially following forest fires. Aspen frequently propagates by suckers from roots. Growth will continue for 80–100 years before the stands begin to deteriorate. Also in forests with white spruce and birch. Rare hybrids with balsam poplar have been noted.

USES The wood has not yet been utilized commercially in quantity in Alaska. Elsewhere it is used for pulpwood, boxes and crates, and excelsior.

NOTE Quaking aspen is fast-growing, and the most widely distributed tree species in North America.

QUAKING ASPEN BARK

QUAKING ASPEN

Black Cottonwood

Populus trichocarpa Torr. & Gray ex Hook.

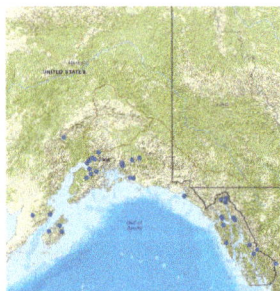

OTHER NAMES balsam cottonwood, Pacific poplar

SYNONYM *Populus balsamifera* subsp. *trichocarpa* (Torr. & Gray) Brayshaw

DESCRIPTION Large deciduous tree to 80–100 feet (24–30 m) tall, with straight trunk 3 feet (1 m) in diameter, with narrow pointed crown; in age larger and developing a tall massive trunk and small flat-topped crown.

LEAVES with slender petioles 1½–2 in. (4–5 cm) long, round, finely hairy. **Leaf blades** broadly ovate, 2½–5 in. (6–12.5 cm) long, 1½–3 in. (4–7.5 cm) wide, mostly long-pointed at tip, rounded or slightly notched at base, with many small rounded teeth, hairless or nearly so, shiny dark green above, beneath whitish and often with rusty specks.

TWIGS red-brown and hairy when young, with orange dots (lenticels); becoming dark gray, sometimes angled, with raised leaf scars showing 3 dots. **Winter buds** large, to 3/4 in. (2 cm) long, long-pointed, sticky or resinous, covered with shiny brown scales.

BARK gray to dark gray, smooth, becoming rough, thick, deeply furrowed with flat ridges.

WOOD with thin whitish sapwood and light brown heartwood, fine-textured, lightweight, soft.

FLOWER CLUSTERS (catkins) 1½–3 in. (4–7.5 cm) long, narrow, drooping, with many small flowers about 1/8 in. (3 mm) long, each with disk and above a light brown hairy lobed scale, male and female on different trees (dioecious). **Male flowers** with 40–60 reddish purple stamens; **female flowers** with rounded densely hairy ovary and 3 broad lobed stigmas. Seed capsules in catkins to 6 in. (15 cm) long, short-stalked, rounded, 3/16 in. (5 mm) in diameter, white hairy, 3-parted, with many tiny cottony seeds.

FLOWERING in May before the leaves, fruit maturing in June–July.

HABITAT found in lowlands of the coastal forests of southeast and southern Alaska. It is best developed at lower levels on river bottoms and sandbars, forming pure stands with undergrowth of willows and alders. It is common on the valley floors of a few large streams, such as Stikine and Taku Rivers; very rare on islands.

USES Trees are planted for shade in towns of southeast Alaska. Southward, the wood is used for boxes and crates, pulpwood, and excelsior. The small supply in Alaska is a possible source of paper pulp, veneer, and lumber. Square cut logs have been used for cabins.

NOTE Black cottonwood is the largest broadleaf tree in Alaska,

BLACK COTTONWOOD

growing rapidly to a height of 80–100 feet (24–30 m) at maturity. It is also the hardwood or broadleaf tree of greatest size in northwestern North America, reaching a height of 125 feet (38 m) on the best sites at age 35 years.

Black cottonwood is not easily distinguished from its close relative, **balsam poplar**. Both have much the same general appearance and similar habitats. The chief differences are in the seed capsules, which in black cottonwood are nearly round, densely hairy, and split into 3 parts, and which in balsam poplar are longer than broad and long-pointed, hairless but warty, and split into 2 parts. Also, there are minor differences in flowers. The pistil of black cottonwood has 3 carpels and 3 stigmas, while that of balsam poplar has 2 carpels and 2 stigmas. The number of stamens is reported to be greater in black cottonwood. Leaves of black cottonwood generally are wider in proportion to length and seem to be whiter beneath.

Black cottonwood hybridizes extensively with balsam poplar where the ranges meet and overlap slightly, for example, in the Cook Inlet and Lynn Canal areas. Hybrids or intermediate trees are recognized by the seed capsules, which may be 3-parted and hairless or 2-parted and hairy. As the ranges of the two species are mostly separate, most trees or specimens without seed capsules can be identified by their locality.

WILLOW *Salix*

Willows are well represented in Alaska, as in other far northern lands. In habit they vary from prostrate or creeping dwarf shrubs, to erect bushes 2–6 feet (0.6–2 m) tall, to large shrubs or small trees, usually with many stems. As many variations occur, and some species seem to intergrade or hybridize, correct identification can be difficult, and a number of additional varieties or subspecies of Alaska *Salix* have been proposed.

Although field identification can be difficult, especially in winter, the willows as a group can be distinguished by: the usually slender or wiry twigs, the winter buds covered by a single

BLACK COTTONWOOD

bud-scale, and by the bitter, quinine-like taste of the bark. The leaves are short-stalked, generally long and narrow, with smooth or finely toothed edges. The yellowish or greenish male and female flowers are borne in hairy, narrow catkins 1–3 in. (2.5–7.5 cm) long, on separate plants in early spring, before or with the leaves. The fruits in the catkins are pointed, thin-walled seed capsules about 1/4 in. (6 mm) long, which split open in spring and summer to release numerous tiny seeds tufted with cottony hairs.

Shrubby willows are widely distributed almost throughout Alaska, extending beyond the limits of trees to the Arctic coast, Bering Sea, and Aleutian Islands. They dominate the undergrowth of the open spruce-birch forest of interior Alaska, and form thickets on sandbars and other porous soils along streams. Although not suitable for lumber because of their small size, shrubby willows provide important summer and winter food for many game animals, especially moose and ptarmigan.

The great variation in willows makes it difficult to construct a completely workable key, and two keys are provided. The **first key** uses features of both the leaves and mature catkins, and material to be identified should be growing under normal conditions (not sprouts or fast-growing roadside shoots). Because the catkins often develop before the leaves, it may be necessary to tag the plant and return to it later in the growing season. In addition, unusual growth forms resulting from differences in site conditions cannot always be incorporated into the key. For example, a tall-growing shrub may become low and prostrate near its range limits. The **second key**, for specimens without catkins, is a vegetative key based on the willow's growth form, leaves, and twigs. Note that this key omits a number of the less common willows reported from Alaska.

With experience, one can learn to distinguish many willows. It is easiest to start with the more common and distinctive willows, such as Sitka willow and Scouler willow in southeastern Alaska, and feltleaf willow, diamondleaf willow, and Bebb willow in central Alaska.

ADDITIONAL SPECIES (*included in first key (noted with an *) but not further described*)
Coin-leaf Willow (*Salix nummularia* Anderss.), known in Alaska only from St. Paul Island in the Pribilof Islands; widespread in Eurasia.
Planeleaf Willow (*Salix planifolia* Pursh), in Alaska, only known from Skagway area.
Mackenzie's Willow (*Salix prolixa* Anderss.) reported from southeast Alaska.

1. KEY TO ALASKA WILLOWS

NOTE *Key adapted from Argus 2004; measurements given in metric units only.*

1	Dwarf or prostrate trailing shrubs under 2 dm tall	2
1	Erect shrubs, greater than 2 dm tall, or trees	20

Dwarf or prostrate trailing shrubs under 8 inches (2 dm) tall

2	Leaves prominently net-veined on upper surface and pale on undersurface; catkins terminal on previous year's branches	NETLEAF WILLOW (*Salix reticulata, p. 216*)
2	Leaves not as prominently net-veined; catkins borne laterally on previous year's branches	3
3	Ovaries hairy (sometimes only the beak hairy)	4
3	Ovaries glabrous	13
4	Leaf margins distinctly serrulate	CHAMISSO WILLOW (*Salix chamissonis, p. 200*)
4	Leaves entire, or toothed on lower half	5
5	Leaf underside not glaucous	6
5	Leaf underside glaucous	7
6	Branches with persistent, skeletonized leaves; leaf margins usually ciliate	SKELETONLEAF WILLOW (*Salix phlebophylla, p. 212*)
6	Branches lacking persistent leaves; leaf margins rarely ciliate	POLAR WILLOW (*Salix polaris, p. 213*)

7 Leaves 8–14 mm long; margins prominently ciliate; catkins usually globose
OVALLEAF WILLOW (*Salix ovalifolia, p. 211*)

7 Leaves longer than 14 mm; margins not ciliate, catkins usually stout to slender   **8**

8 Styles short, to only 0.5 mm long   **9**

8 Styles longer than 0.5 mm   **10**

9 Leaves mostly obovate to elliptic, 1.4–2.5 times longer than wide, upper surface glabrous; margins distinctly toothed on lower half; ovaries sparsely hairy with rust-colored hairs
ALASKA BOG WILLOW (*Salix fuscescens, p. 202*)

9 Leaves mostly narrowly elliptic to narrowly obovate, 1.6–3.8 times longer than wide, hairy on both surfaces; margins entire; ovaries densely white hairy
BARREN-GROUND WILLOW (*Salix niphoclada, p. 210*)

10 Ovaries sparsely hairy with crinkled hairs; branchlets slender and trailing, glabrous; leaves glabrous   EASTERN ARCTIC WILLOW (*Salix arctophila, p. 190*)

10 Ovaries sparsely or densely hairy; branchlets and leaves various   **11**

11 Ovaries usually densely hairy; leaves dark green and usually shiny on upper surface; branchlets trailing to erect   ARCTIC WILLOW (*Salix arctica, p. 190*)

11 Ovaries glabrous to sparsely hairy on beak; leaves shiny to glossy on upper surface; branchlets trailing and rooting   **12**

12 Leaf bases usually cuneate, sometimes acute, upper leaf surface shiny
WEDGELEAF WILLOW (*Salix sphenophylla, p. 190*)

12 Leaf bases acute to rounded or subcordate, upper leaf surface glossy
CREEPING WILLOW (*Salix stolonifera, p. 225*)

13 Leaf underside green (not glaucous)   **14**

13 Leaf underside glaucous   **16**

14 Decumbent or trailing boreal shrub; leaves narrowly elliptic to narrowly obovate, 17–74 mm long; margins crenate to crenate-serrulate
LOW BLUEBERRY WILLOW (*Salix myrtillifolia, p. 209*)

14 Dwarf, sometimes trailing arctic shrubs; leaves circular to obovate or narrowly elliptic, 4–22 mm long; margins entire or toothed only at base   **15**

15 Leaves usually subcircular, somewhat net-veined on upper surface; margins glandular-toothed on lower half, not ciliate; branches more or less trailing
COIN-LEAF WILLOW (*Salix nummularia, p. 180*)*

15 Leaves circular or sometimes broadly elliptic, not prominently reticulate; margins entire and ciliate; branches erect, not trailing   LEAST WILLOW (*Salix rotundifolia, p. 219*)

16 Branchlets usually densely woolly; leaves lemon green, leathery, obovate to narrowly obovate and tapering to a short petiole less than 0.5 mm long; ovaries brick red
SETCHELL WILLOW (*Salix setchelliana, p. 222*)

16 Branchlets glabrous to sparsely hairy; leaves thin, elliptic to subcircular, petioles 1–2.5 mm long; bracts brown to blackish; ovaries reddish, purple, or greenish   **17**

17 Leaf margins distinctly toothed on lower half, petioles 2–5.5 mm long
ALASKA BOG WILLOW (*Salix fuscescens, p. 202*)

17 Leaf margins usually entire, petioles usually 4–20 mm long   **18**

18 Leaf bases usually wedge-shaped, upper leaf surface shiny
WEDGELEAF WILLOW (*Salix sphenophylla, p. 190*)

18 Leaf bases acute to rounded or subcordate, upper leaf surface glossy   **19**

19 Branches short and erect, sometimes trailing, often glaucous; plants often rhizomatous; styles 0.5–1.5 mm long   CREEPING WILLOW (*Salix stolonifera, p. 225*)

19 Branches usually long and trailing, not glaucous; styles to 1 mm long
OVALLEAF WILLOW (*Salix ovalifolia, p. 211*)

Erect shrubs, greater than 8 inches (2 dm tall), or trees

20 Flowering before leaves emerge, catkins not on distinct leafy branchlets 21

20 Flowering with or just before the leaves emerge, catkins on distinct leafy branchlets 29

21 Ovaries glabrous 22

21 Ovaries hairy 25

22 Stipules absent; branchlets brittle at base and with persistent, long, villous hairs at base
COASTAL WILLOW (*Salix hookeriana, p. 206*)

22 Stipules leaflike, often persistent; branchlets flexible, lacking long, villous hairs at base
23

23 Stipules linear to ovate, persisting for several years; styles 1.5–3 mm long
RICHARDSON'S WILLOW (*Salix richardsonii, p. 218*)

23 Stipules broadly ovate, not persisting for more than one year; styles usually shorter than 1.5 mm 24

24 Catkins flowering before leaves emerge; flowering branchlets to 5 mm long; styles 0.5–2 mm long; branchlets often glabrous or sparsely hairy; leaves broadly elliptic to obovate
PARK WILLOW (*Salix pseudomonticola, p. 214*)

24 Catkins flowering with, or sometimes just before, leaves emerge; flowering branchlets 0.5–6 mm long; styles to 1 mm long; branchlets glabrescent to villous; leaves narrowly oblong to obovate MACKENZIE'S WILLOW (*Salix prolixa, p. 180*)*

25 Leaves densely white woolly on undersurface, bright green on upper surface; stipes to 0.5 mm long FELTLEAF WILLOW (*Salix alaxensis, p. 187*)

25 Leaves silky or densely villous to glabrescent on undersurface; stipes to 3 mm long 26

26 Buds and stipules oily; stipules broadly ovate; margins prominently glandular; leaves white, silky woolly on underside BARRATT WILLOW (*Salix barrattiana, p. 195*)

26 Buds and stipules not oily; stipules ovate to linear; leaves glabrous or silky to glabrescent on underside 27

27 Branchlets velvety-hairy; styles to 0.5 mm long; leaves often oblanceolate, undersurface with scattered white and rust-colored hairs SCOULER WILLOW (*Salix scouleriana, p. 221*)

27 Branchlets pubescent or villous to glabrescent; styles to 2 mm long; leaves usually elliptic to narrowly elliptic, oblong, or rhomboid 28

28 Stipules oblong, narrowly elliptic, or ovate, often rudimentary, and rarely persistent for more than one year, 1–3 mm long, shorter than the petioles; young leaves usually more hairy; mature leaves usually narrowly ellipticPLANELEAF WILLOW (*Salix planifolia, p. 180*)*

28 Stipules linear, often persistent for two to four years, 3–30 mm long, longer than petioles; young leaves glabrous or sparsely pilose; mature leaves often rhombic, or narrowly elliptic to obovate TEA-LEAF WILLOW (*Salix pulchra, p. 215*)

29 Ovaries glabrous 30

29 Ovaries hairy 39

30 Leaf underside green or pale, not glaucous 31

30 Leaf underside glaucous 35

31 Petioles with glandular dots or lobes near their base
PACIFIC WILLOW (*Salix lasiandra, p. 208*)

31 Petioles lacking glandular dots of lobes near their base 32

32 Leaves linear, 7–19 times longer than wide; margins distantly denticulate; catkins sometimes branched; pistillate bracts deciduous after flowering
SANDBAR WILLOW (*Salix interior, p. 207*)

32 Leaves not linear, only 2–5 times longer than wide; margins serrulate or crenate; catkins not branched; pistillate floral bracts persistent 33

33 Leaves coarsely long-hairy on both surfaces; margins glandular-serrulate or partly entire
UNDERGREEN WILLOW (*Salix commutata, p. 201*)

33 Leaves glabrous or becoming so; margins glandular-crenate to crenate-serrulate   34

34 Shrubs decumbent, mostly 10–60 cm tall; stipules rudimentary, 1–2 mm long; styles to 1 mm long; proximal leaf margins crenulate
LOW BLUEBERRY WILLOW (*Salix myrtillifolia, p. 209*)

34 Shrubs erect, 0.6–7 m tall; stipules foliaceous, 1–5 mm long; styles to 1.5 mm long; lower leaf margins entire or serrulate   TALL BLUEBERRY WILLOW (*Salix boothii, p. 198*)

35 Petioles glandular doted or lobed near their base; stamens 5; leaf tips caudate to acuminate   PACIFIC WILLOW (*Salix lasiandra, p. 208*)

35 Petioles not glandular near their base; stamens 2; leaf tips acute to rounded   36

36 Stipules rudimentary or absent   COASTAL WILLOW (*Salix hookeriana, p. 206*)

36 Stipules leaflike   37

37 Styles to 0.5 mm long; leaves with tiny rust-colored hairs on upper surface midrib
HALBERD WILLOW (*Salix hastata, p. 205*)

37 Styles 0.5–1.5 mm long; leaves with white hairs, if any   38

38 Leaves elliptic or obovate; young leaves green and opaque; petioles green; branchlets densely to sparsely long-hairy   BARCLAY WILLOW (*Salix barclayi, p. 193*)

38 Leaves narrowly oblong to narrowly obovate; juvenile leaves reddish and translucent; petioles reddish; branches glabrescent   MACKENZIE'S WILLOW (*Salix prolixa, p. 180*)*

39 Flowering just before leaves emerge, staminate catkins appearing just before leaves, flowering branchlets to 8 mm long; pistillate catkins appearing after leaves, flowering branchlets 1-7 mm long; stipules absent or rudimentary; styles to 0.5 mm long; stipes to 5 mm long   BEBB WILLOW (*Salix bebbiana, p. 196*)

39 Plants without this set of above characters   40

40 Stipes 3–5 mm long   BEBB WILLOW (*Salix bebbiana, p. 196*)

40 Stipes to 2 mm long   41

41 Leaf undersurface silky; margins glandular-serrulate to only sparsely so   42

41 Leaf undersurface densely pubescent to glabrescent, not silky   43

42 Leaves very narrowly elliptic to elliptic, 5–7 times longer than wide, undersurface silky with short, white or rust-colored hairs; margins prominently glandular-serrulate
LITTLETREE WILLOW (*Salix arbusculoides, p. 188*)

42 Leaves narrowly elliptic to obovate, 2–3 times longer than wide, appearing satiny beneath with appressed, silky hairs; margins distantly and inconspicuously glandular-serrulate to glandular-crenate   SITKA WILLOW (*Salix sitchensis, p. 224*)

43 Young leaves with white and rust-colored hairs
ATHABASKA WILLOW (*Salix athabascensis, p. 192*)

43 Young leaves with white hairs   44

44 Young leaves very densely tomentose; petioles woolly; mature leaves densely dull white-woolly on underside, floccose to nearly hairless on upper surface; blades 3.5–12 times longer than wide; styles red   SILVER WILLOW (*Salix candida, p. 199*)

44 Young leaves long-silky; petioles glabrate to villous; mature leaves not hairy as in *Salix candida,* blades 1.5–5 times longer than wide; styles yellow-green   45

45 Petioles 3–15 mm long, yellowish; stipes to 2 mm long
GRAYLEAF WILLOW (*Salix glauca, p. 203*)

45 Petioles 1–3 mm long, often reddish; stipes to 0.5 mm long
BARREN-GROUND WILLOW (*Salix niphoclada, p. 210*)

2. VEGETATIVE KEY TO ALASKA WILLOWS

Because leaf, twig, and growth form characteristics of some willows are extremely variable, a vegetative key cannot account for all the variability encountered in the field. The following key treats about three-fourths of Alaska's willows. However, it is inevitable that some plants will key to the wrong species, but it should be possible to narrow the choice to 2 or 3 species, followed by a check of the species descriptions, photographs, and range maps.

1	Low, prostrate shrubs less than 12 in. (30 cm) high	2
1	Erect shrubs or trees, more than 1 foot (30 cm) high	12

Low, prostrate shrubs less than 12 in. (30 cm) high

2 Creeping shrubs with long prostrate branches, often rooting at nodes, but with branches ascending from 4–12 in. (10–30 cm); leaves more than 1 in. (2.5 cm) long   **3**

2 Matted or creeping shrubs, usually less than 4 in. (10 cm) tall, usually in compact mats without long creeping branches; leaves less than 1 in. (2.5 cm) long, entire   **9**

3 Leaves toothed around margin, green on both surfaces or sometimes lighter green beneath   **4**

3 Leaves entire or toothed only on lower half, green above, whitish (glaucous) beneath   **6**

4 Leaves bluish green, leathery or fleshy, 3–4 times longer than wide, tapering gradually to base   **SETCHELL WILLOW** (*Salix setchelliana, p. 222*)

4 Leaves not bluish green, oval, not tapering to base, thin   **5**

5 Leaves nearly as wide as long, elliptic, 3/4–2 in. (2–5 cm) long; branches prostrate   **CHAMISSO WILLOW** (*Salix chamissonis, p. 200*)

5 Leaves 2–3 times longer than wide, 3/8–1½ in. (1–4 cm) long, branches ascending   **LOW BLUEBERRY WILLOW** (*Salix myrtillifolia, p. 209*)

6 Leaves dark green above, conspicuously net-veined, round, with long red petiole   **NETLEAF WILLOW** (*Salix reticulata, p. 216*)

6 Leaves not conspicuously net-veined, more than 2 times as long as broad, petiole green   **7**

7 Leaves fleshy, 3–4 times longer than wide, tapering to base, bluish green; on dry gravel sites   **SETCHELL WILLOW** (*Salix setchelliana, p. 222*)

7 Leaves not fleshy or bluish green, 2 times longer than wide, not tapering to base; in bogs or on arctic and alpine tundra   **8**

8 Trailing shrub with long branches rooting at nodes, leaves finely glandular toothed on basal half; usually in boggy sites   **ALASKA BOG WILLOW** (*Salix fuscescens, p. 202*)

8 Leaves entire, forming dense mats from short branches; mostly in dry alpine and arctic sites   **ARCTIC WILLOW** (*Salix arctica, p. 190*)

9 Leaves green on both surfaces   **10**

9 Leaves green above, whitish (glaucous) beneath   **OVALLEAF WILLOW AND CREEPING WILLOW** (*Salix ovalifolia* and *S. stolonifera, p. 211 and 225*)

10 Shrubs densely matted, often from a central taproot; leaves less than 3/4 in. (2 cm) long; stems brown to reddish brown   **11**

10 Shrubs forming loose mats, usually with long trailing buried branches; stems pale yellow, thin; leaves to 1 in. (2.5 cm) long, usually smaller   **POLAR WILLOW** (*Salix polaris, p. 213*)

11 Shrub mat with abundant dead leaves persistent; leaves 3/8–3/4 in. (1–2 cm) long   **SKELETONLEAF WILLOW** (*Salix phlebophylla, p. 212*)

11 Shrubs with few or no dead leaves; leaves 1/8–3/8 in. (4–10 mm) long   **LEAST WILLOW** (*Salix rotundifolia, p. 219*)

Erect shrubs or trees, more than 1 foot (30 cm) high

12 Upright shrubs usually less than 3 feet (1 m) high   13
12 Tall shrubs or trees 3–25 feet (1–7.5 m) or more in height   20
13 Leaves with hairs on lower surface, gray or silvery   14
13 Leaves without conspicuous hairs   16
14 Leaves linear to lanceolate, 5–7 times longer than wide, with dense woolly hairs beneath; rare shrub of interior bogs   SILVER WILLOW (*Salix candida, p. 199*)
14 Leaves broader, not densely woolly beneath   15
15 Leaves with dense straight hairs, often oriented in vertical plane; petioles green, yellow, or brown; low compact shrub with thick branches; bud scales giving off yellow waxy substance when plant is dried   BARRATT WILLOW (*Salix barrattiana, p. 195*)
15 Leaves with scattered hairs; petioles reddish; upright shrub with slender branches; buds not giving off waxy substance   BARREN-GROUND WILLOW (*Salix niphoclada, p. 210*)
16 Leaves fleshy, bluish green, 3–4 times longer than wide, tapering gradually to base   SETCHELL WILLOW (*Salix setchelliana, p. 222*)
16 Leaves thin, green, oval   17
17 Stipules, if present, persisting less than 1 year.
17 Stipules persistent for several years   19
18 Leaves toothed around margin, lower surface light green, not whitish (glaucous)   LOW BLUEBERRY WILLOW (*Salix myrtillifolia, p. 209*)
18 Leaves toothed only on basal half with fine glandular teeth, lower surface whitish (glaucous)   ALASKA BOG WILLOW (*Salix fuscescens, p. 202*)
19 Stipules broad at base and glandular toothed along margins; twigs coarse, brown to black, with dense hairs persistent for several years   RICHARDSON'S WILLOW (*Salix richardsonii, p. 218*)
19 Stipules linear, narrow at base, without glandular teeth; twigs fine, usually reddish brown and shiny, without dense hairs after 1 year   TEA-LEAF WILLOW (*Salix pulchra, p. 215*)
20 Leaves linear, 1½–4 in. (4–10 cm) long, and 1/4 in. (6 mm) wide, with scattered small teeth; usually growing on river alluvium   SANDBAR WILLOW (*Salix interior, p. 207*)
20 Leaves broader   21
21 Adult leaves with hairs on lower surface   22
21 Adult leaves without hairs on lower surface   31
22 Lower surface of leaves with dense hairs, appearing silvery, white, or gray   23
22 Lower surface of leaves visible through less dense hairs   26
23 Lower surface of leaves with dense white *woolly* hairs   24
23 Lower surface of leaves with dense *straight* hairs.   25
24 Leaves long and narrow, lance-shaped, 2–4 in. (5–10 cm) long and only 1/4–5/8 in. (6–15 mm) wide; low shrubs seldom exceeding 4 feet (1.2 m) in height; rare in boggy sites in interior Alaska   SILVERY WILLOW (*Salix candida, p. 199*)
24 Leaves broader, 2–4 in. (5–10 cm) long, and ½–1½ in. (12–40 mm) wide; tall shrub or tree to 30 feet (9 m), common in many sites over most of Alaska   FELTLEAF WILLOW (*Salix alaxensis, p. 187*)
25 Lower surface silky hairy, upper surface green, with scattered hairs; tall shrub or tree to 20 feet (6 m) high   SITKA WILLOW (*Salix sitchensis, p. 224*)
25 Lower surface dull gray hairy, upper surface greenish gray, without hairs; shrub usually less than 10 feet (3 m) high   GRAYLEAF WILLOW (*Salix glauca, p. 203*)
26 Margins of leaves distinctly toothed   27
26 Margins of leaves not toothed or with a few teeth on basal half   28
27 Leaves light green on both surfaces, not shiny, oval, about 2 times longer than wide; shrub 3–6 feet (1–2 m) high   UNDERGREEN WILLOW (*Salix commutata, p. 201*)

27 Leaves dark green and shiny above, whitish (glaucous) beneath, 3–4 times longer than wide; shrub 10–15 feet (3–4.5 m) tall, with slender branches
LITTLETREE WILLOW (*Salix arbusculoides, p. 188*

28 Hairs on lower surface short and stiff, at least some red, giving a reddish hue
SCOULER WILLOW (*Salix scouleriana, p. 221*)

28 Hairs denser, longer, not reddish   29

29 Tall shrubs or trees 10–25 feet (3–7.5 m) tall; twigs diverging at nearly right angles from the main stem   BEBB WILLOW (*Salix bebbiana, p. 196*)

29 Medium shrubs, usually under 10 feet (3 m) high; twigs usually branching at 45° angle or less   30

30 Petioles 1/8–3/8 in. (3–10 mm) long, yellow, leaves obovate to oblong, acute to obtuse
GRAYLEAF WILLOW (*Salix glauca, p. 203*)

30 Petioles less than 1/8 in. (3 mm) long, reddish, leaves strap-shaped, rounded or blunt
BARREN-GROUND WILLOW (*Salix niphoclada, p. 210*)

31 Stipules persistent on the twigs several years   32

31 Stipules not persisting more than 1 year   33

32 Stipules broad at the base and glandular toothed along margins; twigs coarse, brown to black, densely hairy, even after several years RICHARDSON'S WILLOW (*S. richardsonii, p. 218*)

32 Stipules linear, narrow at base, without glandular teeth; twigs, fine, usually reddish brown, shiny, without dense hairs at 1 year   TEA-LEAF WILLOW (*Salix pulchra, p. 215*)

33 Leaves with teeth around margin   34

33 Leaves entire or with teeth only on lower part   38

34 Leaves 3–4 times as long as broad; tall shrubs or trees   35

34 Leaves less than 3 times as long as broad   36

35 Leaves large, 3–4 in. (7.5–10 cm) long, lance-shaped, with long tapering tip; young twigs woolly   PACIFIC WILLOW (*Salix lasiandra, p. 208*)

35 Leaves smaller, 2–3 in. (5–8 cm) long, not lance-shaped, short-pointed; young twigs not woolly   LITTLETREE WILLOW (*Salix arbusculoides, p. 188*)

36 Leaves whitish (glaucous) beneath   37

36 Leaves light green, not whitish (glaucous) beneath
TALL BLUEBERRY WILLOW (*Salix boothii, p. 198*)

37 Leaves broadly lance-shaped to oval, usually narrowing to small projection at tip (apiculate), often reddish when young; well drained alluvial soils and upland forests
PARK WILLOW (*Salix pseudomonticola, p. 214*)

37 Leaves ovate, blunt at tip; not reddish when young; moist habitats in open and forested areas and near tree-line   BARCLAY WILLOW (*Salix barclayi, p. 193*)

38 Tall shrubs or trees, 10–25 feet (3–7.5 m) tall   39

38 Smaller shrubs, to 10 ft (3 m) tall, occasionally taller   40

39 Leaves large at maturity, to 3 in. (7.5 cm) long, obovate; plants mostly in Yakutat Bay area
COASTAL WILLOW (*Salix hookeriana, p. 206*)

39 Leaves smaller, 1½–2½ in. (4–6 cm) long, dull grayish green above, elliptic to ovate; widely distributed in most of Alaska except southeast BEBB WILLOW (*S. bebbiana, p. 196*)

40 Leaves strap-shaped, grayish; petioles reddish, stipules absent
BARREN-GROUND WILLOW (*Salix niphoclada, p. 210*)

40 Leaves oval, green, petioles green or yellow, stipules usually present   41

41 Upper leaf surface light green, not shiny; typically on river alluvium, interior and northern Alaska   HALBERD WILLOW (*Salix hastata, p. 205*)

41 Upper leaf surface dark green, usually shiny; typically in moist sites in open and forested areas in coastal Alaska   BARCLAY WILLOW (*Salix barclayi, p. 193*)

Feltleaf Willow

Salix alaxensis (Anderss.) Coville

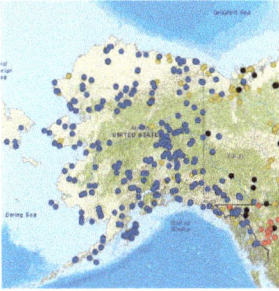

SYNONYMS *Salix longistylis* Rydb.

DESCRIPTION A shrub or small tree to 20–30 feet (6–9 m) high with a trunk 4–7 in. (10–18 cm) in diameter, occasionally dwarfed and nearly prostrate in exposed places.

LEAVES elliptic or oblanceolate (reverse lance-shaped), 2–4 in. (5–10 cm) long and ½–1½ in. (1.2–4 cm) wide, short-pointed, usually tapering to base, edges without teeth or nearly so, above dull green and hairless or sometimes somewhat short-hairy, beneath covered with a dense white or creamy-white felt; midrib yellowish.

TWIGS One-year and 2–year twigs stoutish, usually white-woolly.

BARK gray, smooth, becoming rough and furrowed into scaly plates.

CATKINS stoutish, not stalked, appearing before the leaves, 2–4 in. (5–10 cm) long at maturity; scales blackish.

SEED CAPSULES long, pointed, white-woolly.

FLOWERING May and June, seeds ripening in June and July. The leaves appear in early May and are fully developed by mid-June.

HABITAT Widely distributed in valleys almost throughout Alaska. Extending beyond the limits of the spruce-birch interior forest, it is the only tree-sized willow in many areas.

USES In many places in northern Alaska, this willow is important as the only wood available for fuel. Though not the common "diamond willow" from which ornamental canes are made, the trunks sometimes have this pattern of diamond-shaped scars where the lower twigs have died.

Feltleaf willow is a preferred browse species of moose which pull down and break branches and trunks up to 1½ in. (4 cm) in diameter. Eventually the shrub grows above the reach of the moose and the stems become too thick for the moose to break. It is reported that the inner bark has served as emergency food for humans.

NOTE This willow was first collected at Kotzebue Sound beyond Bering Strait. The specific name *alaxensis* means Alaskan but is from an old spelling.

TIP *Salix candida* is the only other willow species with leaf underside densely covered with woolly hairs. *Salix candida,* uncommon in Alaska, differs from *S. alaxensis* by the catkins borne on leafy branchlets, and by the elongated leaves less than 3/4 in. (2 cm) wide.

FELTLEAF WILLOW

FELTLEAF WILLOW

Littletree Willow
Salix arbusculoides Anderss.

DESCRIPTION An erect shrub 10–15 feet (3–4.5 m) tall or commonly a small tree 25–30 feet (7.5–9 m) tall and 5–6 in. (12–15 cm) in trunk diameter.

LEAVES narrowly elliptic-lanceolate, often oblanceolate while unfolding, 1–3 in. (2.5–7.5 cm) long, 3/8–3/4 in. (10–20 mm) wide, usually short-pointed at both ends, with margins finely shallowly toothed, green and hairless above, beneath whitish to white and finely silvery-hairy; veins closely parallel.

TWIGS slender, much branched, the younger yellowish brown and sometimes thinly short-hairy, the older reddish brown, hairless, and shiny.

BARK gray to reddish brown, smooth.

CATKINS small and slender on very short stalks, appearing slightly before or with the leaves, 1–2 in. (2.5–5 cm) long at maturity; scales blackish.

SEED CAPSULES small, thinly silvery-hairy.

FLOWERING mid-May to early June, seeds ripening mid-to late June. Catkins develop before or at the same time as the leaves, which are fully developed as early as the end of May.

HABITAT A common willow, forming dense thickets along streams and rivers in interior Alaska. It also grows in uplands along streams and is a common successional species following the burning of open, wet stands of black spruce. It is less commonly found as a shrub in stands of white spruce and birch. On the north slope of the Brooks Range, it grows on streambanks and gravel bars together with several other willow species.

NOTE One of several species that form "diamond willow" patterns.

TIP Alaska's only willow with the combination of leaves more than 3 times longer than wide, leaves silky-hairy on underside, and the leaf margins finely glandular-toothed. *Salix pulchra* may have elongated leaves, but the leaf underside is hairless, the leaf margins are not toothed, and the stipules remain on the stems for several years.

LITTLETREE WILLOW

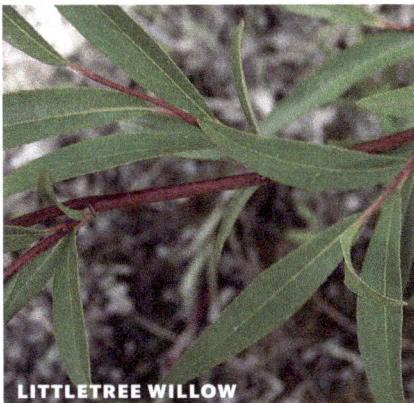

LITTLETREE WILLOW

Arctic Willow
Salix arctica Pallas

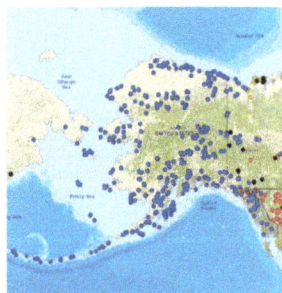

SYNONYMS *Salix angolorum* Cham., *Salix crassijulis* Trautv., *Salix torulosa* Trautv.

DESCRIPTION A trailing low shrub frequently forming dense mats to 8–10 in. (20–25.5 cm) high, commonly lower.

LEAVES variable in shape but generally obovate to elliptic 3/4–3 in. (2–7.5 cm) long and 3/8–1¼ in. (1–3 cm) wide, blunt or short-pointed at tip. Upper surface dark green and often shiny, under surface pale green, margins entire. Petioles 3/16–5/8 in. (5–15 mm) long.

TWIGS reddish, coarse, and much branched, rooting at nodes.

CATKINS relatively large, to 4 in. (10 cm) long and 5/8 in. (15 mm) thick, erect on stalks that may be leafless or with 2 or 3 leaves; scales brown to black with long silky hairs.

SEED CAPSULES broad, 5/16 in. (8 mm) long, with scattered hairs, reddish to pale brown.

FLOWERING in June and July, fruits ripening in July and August. Catkins develop at the same time as the leaves.

HABITAT In arctic and alpine tundra, Arctic willow may occur as a loose trailing shrub or compact low mats. It is found in both dry and wet sites and in protected and exposed situations. In southeast Alaska, although primarily in the alpine tundra, it may occur at sea level on glacial outwash and moraines.

NOTE Arctic willow is extremely variable in growth form and in the size and shape of the leaves.

TIP In *Salix arctica,* the leaf undersurface is usually covered with whitish hairs, and usually at least some leaves of have long white hairs on their undersides and margins, forming a 'beard' at the tip; in *S. rotundifolia, S. phlebophylla,* and *S. polaris,* the leaves are hairless, and the undersurface is green. The ovaries in *Salix stolonifera* and *S. ovalifolia* are hairless. Plants of *S. arctica* that lack the 'beard' hair at the leaf tip and have no female catkins may be difficult to distinguish from *S. stolonifera* and *S. ovalifolia.*

ADDITIONAL SPECIES The closely related **eastern arctic willow** (*Salix arctophila* Cockerell ex Heller) of eastern Canada occurs in northeasten Alaska. It resembles arctic willow (*S. arctica*) in appearance but has darker, more shiny upper leaf surfaces, the leaf underside is hairless, and plants often have long, trailing, yellow-green stems.

Another closely related species, **wedge-leaf willow** (*Salix sphenophylla* A. Skvort.), of northeastern Asia, is known from the Seward Peninsula and far northern Alaska.

ARCTIC WILLOW

ARCTIC WILLOW

ARCTIC WILLOW

Athabasca Willow
Salix athabascensis Raup

SYNONYMS *Salix fallax* Raup, *Salix pedicellaris* var. *athabascensis* (Raup) Boivin

DESCRIPTION Shrub 2–7 feet (0.6–2.1 m) tall, uncommon in interior Alaska.

LEAVES narrowly elliptic to clliptic or narrowly obovate, the largest mature leaves 3/4–2 in. (1.7-5 cm) long and 3/8–5/8 in. (9–15 mm) wide, the tip acute; margins entire and often glandular on lower portion of blade; the leaves more or less covered with white or reddish hairs lying flat against the surface of the leaf; petioles 1/8–3/8 in. (3–9 mm) long; **stipules** tiny, less than 1/32 in. (1 mm) long.

TWIGS reddish brown, glossy, densely or sparsely pubescent, becoming nearly hairless.

CATKINS developing with the leaves; **staminate catkins** 1/4–3/8 in. (5–9 mm) long; **pistillate cakins** 1/2–1½ in. (1.2–3.7 cm) long, loosely flowered, the flower-bearing branchlets to 5/8 in. (1.5 cm) long; densely hairy with whitish hairs.

HABITAT Fens, muskegs, bogs; from central Alaska along the Tanana River, and in southern Yukon; eastward to the Northwest Territories and Hudson Bay, south to British Columbia and Alberta.

NOTE *Salix athabascensis* is reported to form hybrids with *S. pedicellaris*.

TIP May be difficult to identify as the characteristic reddish hairs of the leaves are often sparse or sometimes absent. Distinguished from *Salix niphoclada* by the longer stipes; however, plants without catkins and lacking the reddish hairs may be confused with *Salix niphoclada*.

ATHABASCA WILLOW

Barclay Willow
Salix barclayi Anderss.

DESCRIPTION Spreading, much branched shrubs tending to form dense thickets 3–6 feet (1–2 m) high, sometimes to 10–20 feet (3–6 m) and treelike, variable in growth form, leaf shape, color, and habitat.

LEAVES broadly elliptic to obovate, 3/4–3 in. (2–7.5 cm) long and 3/8–1¼ in. (1–3 cm) wide, short-pointed tip and wedge-shaped to rounded at base, margins toothed, serrate to entire; upper surface shiny yellow green thinly hairy when young but becoming hairless, sometimes with short reddish hairs along midrib; lower surface whitish (glaucous), thinly hairy but soon becoming hairless; usually turning black in drying. Petioles 1/16–1/4 in. (1.5–6 mm) long. **Stipules** inconspicuous and soon dropping.

TWIGS blackish and densely hairy when young, becoming reddish brown and hairless with blackish buds.

BARK gray or greenish brown, smooth.

CATKINS 2–3 in. (5–7.5 cm) long, on stalks with 2–3 leaves, appearing with the leaves, scales about 1/32 in. (1 mm) long, black, with long hairs.

SEED CAPSULES short and stout, on short stalk, hairy when young but soon becoming hairless, green to reddish.

FLOWERING in June, seeds ripening in July, most catkins falling by August.

HABITAT The most common thicket-forming shrub along the southern coast of central Alaska in forest openings, along small streams, and in wet alluvial sites. On the Kenai Peninsula it forms extensive thickets at tree-line in areas where it is protected by winter snow deposits. *Salix barclayi* is a dominant species in subalpine willow thickets, and in moist disturbed sites at lower elevations south of the Alaska Range. In alpine habitats, it reaches higher elevations in moist gullies. *Salix barclayi* is also very common along roads.

BARCLAY WILLOW

NOTE *Salix barclayi* is rare or possibly absent north of the Alaska Range. Reports of specimens of *S. barclayi* north of the Alaska Range are usually mis-identified specimens of *S. hastata,* and close examination will often reveal sparse reddish hairs on the main leaf vein typical of *S. hastata.*

TIP *Salix barclayi* is variable in appearance but can be distinguished from *S. myrtillifolia* by the pale leaf underside, and from *S. richardsonii* and *S. pseudomonticola* by the catkins borne on well-developed leafy branchlets, appearing at the same time as the leaves. In subalpine habitats, *S. barclayi* can be distinguished from *S. richardsonii* by the absence of large dried stipules that persist several years on the stems in *S. richardsonii.*

Closely related to and easily confused with low blueberry willow (*Salix myrtillifolia*) and undergreen willow (*S. commutata*), but these latter willows have leaves green to pale beneath but never whitish (glaucous). Twigs of Barclay willow often end in rounded galls, composed of deformed leaves and caused by insects. Presence of these galls, often called "willow roses," may aid identification.

Where the range of Barclay willow overlaps that of coast willow and undergreen willow, there is considerable difficulty in separating the three species. The following key may help in identification:

1 Leaves green or pale beneath, not whitish   **UNDERGREEN WILLOW** (*Salix commutata*)
1 Leaves whitish (glaucous) beneath   **2**
2 Stipules present; leaves without hairs beneath; lacking long silky hairs at the base of twigs; styles greenish   **BARCLAY WILLOW** (*Salix barclayi*)
2 Stipules absent (sometimes very small on vigorous shoots); leaves long silky hairy beneath, at least along midrib; long silky hairs persistent at the base of twigs; styles red in life, drying dark   **COASTAL WILLOW** (*Salix hookeriana*)

ETYMOLOGY This species honors George Barclay, English botanical collector with the surveying expedition of the ship *Sulphur* along the western coast of America in 1835–41.

BARCLAY WILLOW

Barratt Willow
Salix barrattiana Hook.

DESCRIPTION A low upright shrub, usually 1–2 feet (30–60 cm) tall, commonly forming loose clumps several yards (meters) across.

LEAVES tending to have a vertical orientation, elliptic to obovate, 1½–2½ in. (4–6 cm) long and 1/4 to 1/3 as wide, short-pointed tip, both surfaces grayish from long silky hairs. Petioles to 5/8 in. (15 mm), longest on upper leaves.

TWIGS stout, densely hairy when young and remaining so for many years, older twigs reddish brown to dark brown.

CATKINS 1¼–2 in. (3–5 cm) long, sessile on twigs, erect in habit, appearing in spring before the leaves; scales black, pointed at tip, with long silky hairs.

SEED CAPSULE stout, about 1/4 in. (6 mm) long, with silky white hairs on pedicels 1/16 in. (1.5 mm) long.

HABITAT A rare shrub in Alaska although it may be locally abundant above tree-line on gravel terraces of some rivers in the Alaska Range where it may reach altitudes of 4,600 feet. It also occurs occasionally in wet alpine meadows.

USES Young twigs are browsed by moose, but the twigs and leaves are reported to be very bitter and are avoided by most herbivores.

NOTE Barratt willow is conspicuous among Alaskan willows and easily determined at a distance by its silvery appearance, its low growth forming dense thickets, and its upright leaves and twigs. When collected and pressed, the scales, stipules, and young twigs exude a yellowish oily substance that stains the paper yellow.

ETYMOLOGY Named for Joseph Barratt (1796–1882), American student of willows.

BARRATT WILLOW

BARRATT WILLOW

Bebb Willow
Salix bebbiana Sarg.

OTHER NAMES diamond willow, beak willow

SYNONYMS *Salix rostrata* Richards.

DESCRIPTION A large shrub 10 feet (3 m) tall or a small, bushy tree 15–25 feet (4.5–7.5 m), rarely 35 feet (10.5 m) with trunk diameter of 6–9 in. (15–23 cm).

LEAVES elliptic and pointed at both ends to broadly oblanceolate or obovate-oval and very short-pointed at tip and broad at base, 1–3½ in. (2.5–9 cm) long and 3/8–1 in. (10–25 mm) wide, edges without teeth or somewhat wavy, dull green above, gray or whitish and roughly net-veined beneath, more or less hairy on both sides but becoming less hairy with age.

TWIGS slender, branching at wide angles, yellowish to brown, gray hairy when young but afterward becoming hairless.

BARK gray to dark gray, smooth, becoming rough and furrowed.

WOOD lightweight, brittle.

CATKINS on short leafy stalks, before or with the leaves, at maturity 1–3 in. (2.5–7.5 cm) long and loose, scales narrow, yellowish with reddish tips, hairy.

SEED CAPSULES long, very slender, with short hairs 1/8–3/16 in. (3–5 mm) long, on slender, sparsely hairy stalks.

FLOWERING mid-May through mid-June, seeds ripening by mid-to late June, catkins shed by mid-July, but often a few dried female catkins remain on the shrub over winter. Catkins appear at the same time as the leaves are developing.

HABITAT Bebb willow is the most common upland willow in interior Alaska, occurring as scattered individuals in most forest types. It is also the most common species in the willow stands that follow forest fires on upland sites and in thickets adjacent to streams, swamps, and lakes. In open meadows it forms large spreading shrubs. Widely distributed in interior Alaska, south to the Pacific Coast.

USES It is an important browse species for moose throughout interior Alaska. In winter heavy snows tend to bend the branches down so that they are in reach of both moose and snowshoe hares.

Bebb willow is the most important producer of "diamond willow." This term applies to several species with diamond-shaped patterns on their trunks. When the stems are carved they result in a striking pattern of diamond-shaped cavities with a sharp contrast between the white or cream sapwood and the reddish brown heartwood. Diamond willow is carved into canes, lamp

BEBB WILLOW

posts, furniture and candle holders. In the old roadhouse at Copper Center, the newel posts and balusters of the whole staircase have been carved from diamond willow.

The depressions or "diamonds" are caused by one or more fungi which attack the willow at the junction of a branch with the main trunk. The "diamond willows" occur most commonly under shade of trees or where the site is poor. They are most abundant in the Copper River basin area but occur in Alaska throughout the boreal forest from the Kenai Peninsula northward. In addition to Bebb willow, the following also form "diamonds" although usually to a lesser degree: Park willow, feltleaf willow, littletree willow, and Scouler willow.

In other areas of the United States, Bebb willow formerly was used for baseball bats, charcoal, gunpowder, and withes for furniture and baskets.

NOTE The abundant yellow stamens of the male catkins and the grayish female catkins appear simultaneously, making plants of Bebb willow stand out against the background of the forest.

ETYMOLOGY This species commemorates Michael Schuck Bebb (1833–95), American specialist on willows.

BEBB WILLOW

BEBB WILLOW - CATKIN

Tall Blueberry Willow
Salix boothii Dorn

SYNONYMS *Salix novae-angliae* Anderss., *Salix pseudocordata* Anderss., *Salix myrtillifolia* Anderss. var. *pseudomyrsinites* (Anderss.) Ball, *Salix pseudomyrsinites* Anderss.

DESCRIPTION A tall erect shrub usually 6–8 feet (2–2.5 m) tall, occasionally to 20 feet (6 m) and treelike.

LEAVES elliptic to obovate, 1–3 in. (2.5–7.5 cm) long and about 1/3 as wide, blunt to short-pointed at tip; margins with teeth often glandular tipped, upper surface dark green, lower surface slightly lighter but not whitish (glaucous), prominently net-veined, with long silky hairs when young, soon becoming hairless. Petioles 1/16–1/4 in. (1.5–6 mm) long. **Stipules** variable, small and inconspicuous or to 3/16 in. (5 mm) long, glandular toothed.

TWIGS brown, usually straight, coarse, with dense white silky hairs when young.

CATKINS 3/4–2½ in. (2–6 cm) long on leafy stalks, appearing after the leaves have started to develop; scales short, brown, with long gray hairs.

SEED CAPSULE green to brown, hairless 1/4–5/16 in. (6-8 mm) long on stalks 1/16–1/8 in. (1.5-3 mm) long.

FLOWERING in early to mid-June, seeds maturing in late June to mid-July, catkins falling in late July.

HABITAT A common willow on the silt and sandbars of the Tanana and Yukon Rivers, where it occurs as a pioneer with other willows; also common in willow thickets along small streams and roadsides.

NOTE Closely related to **low blueberry willow** (*Salix myrtillifolia*), and sometimes treated as a subspecies. However, *S. myrtillifolia* is smaller, has hairless leaves, and grows in fens and bogs.

TALL BLUEBERRY WILLOW

TALL BLUEBERRY WILLOW

Silver Willow

Salix candida Flueggé ex Willd.

OTHER NAMES sage willow, hoary willow

DESCRIPTION An erect shrub usually 6 feet (2 m) or less in height, with an overall silvery appearance.

LEAVES oblong to lanceolate, short-pointed at both ends, 2–4 in. (5–10 cm) long and only 1/4–5/8 in. (6–15 mm) wide, edges entire or wavy and commonly rolled toward lower surface; upper surface silvery from dense woolly hairs when young but becoming hairless and dark green with age; lower surface remaining silvery with dense woolly hairs.

TWIGS covered with white woolly hairs when young but becoming smooth and reddish with age.

CATKINS 3/4–2 in. (2–5 cm) long, narrowly cylindrical, stalkless on twigs, in early spring mostly developing with the leaves; scales brown, rounded at tip, with long white hairs.

SEED CAPSULES stout, 1/4 in. (6 mm) long, covered with short dense woolly hairs.

HABITAT An uncommon shrub in Alaska, collected only a few times in bogs and other wet sites, mainly along the Tanana and Yukon Rivers.

NOTE The silvery appearance of leaves, twigs, and catkins, and the narrow leaf shape make silver willow quite distinctive and easily recognized.

TIP Catkins of *Salix candida* appear at the same time as the leaves and are usually borne on a leafy branchlet, while catkins of *S. alaxensis* appear before the leaves and are borne directly on the stem.

SILVER WILLOW

SILVER WILLOW

Chamisso Willow
Salix chamissonis Anderss.

DESCRIPTION Prostrate loosely branched creeping shrub rooting along the branches and 4–6 in. (10–15 cm) high at the ends.

LEAVES broadly obovate, rounded at tip (sometimes abruptly pointed) and wedge-shaped at base, 3/4–2 in. (2–5 cm) long and 2/3–3/4 as wide, glandular toothed on margin; green on both surfaces but slightly paler beneath. Petioles long and slender, 3/8–5/8 in. (10–15 mm) long.

TWIGS gray or brown, coarse, often buried in mosses and rooting at nodes.

CATKINS erect on leafy twigs, about 1½ in. (4 cm) long, developing with the leaves; scales black with grayish hairs.

SEED CAPSULES long and slender with gray hairs, stalkless or on a very short stalk.

FLOWERING from late June through July, seeds ripening July and August. Catkins develop at the same time as the leaves.

HABITAT Arctic tundra of northern and western Alaska and the alpine tundra of interior Alaska. In the Arctic it grows as a very loose creeping shrub in wet meadows, seepage areas, and adjacent to snow fields. It is abundant in the Eagle Summit area north of Fairbanks where it forms loose mats in similar habitats.

TIP Distinguished from the other creeping willows by its glandular toothed leaves and the slender gray hairy capsules. The leaves of *S. fuscescens* are only toothed at the lower half.

ETYMOLOGY Named for Ludolf Adalbert von Chamisso (1781–1838), German botanist who visited Alaska in 1816 and 1817 on the ship *Rurik.*

CHAMISSO WILLOW

CHAMISSO WILLOW

Undergreen Willow
Salix commutata Bebb

SYNONYM *Salix barclayi* var. *commutata* (Bebb) Kelso

DESCRIPTION A much branched dense shrub 3–6 feet (1–2 m) tall, with an overall light green appearance.

LEAVES elliptic to obovate, to 2½ in. (6 cm) long and about 1/3 to 1/2 as wide; entire or glandular toothed on margins, dense gray hairy on both surfaces when young but only thinly hairy with age, light green on both surfaces. Petioles 1/8–1/4 in. (3–6 mm) long. **Stipules** well-developed and leaflike with glandular margins to 3/8 in. (1 cm) long, persistent or deciduous.

TWIGS densely gray hairy when young but becoming hairless with age, dark brown. Buds of next season's catkins are often large and red by mid to late summer.

CATKINS ¾–1½ in. (2–4 cm) long on leafy shoots, developing with or after the leaves; scales brown with dense woolly hairs.

SEED CAPSULES 1/4 in. (6 mm) long, hairless, reddish but becoming brown with age.

FLOWERING mid-June to July, seeds ripening late July and August.

HABITAT Forming thickets in the mountains of south central Alaska at and just above tree-line along small streams and in areas protected by winter snow accumulation, usually with several other willows. It also occurs occasionally along the coast in wet open habitats.

NOTE Undergreen willow is quite similar in appearance and often grows with Barclay willow but the soft green color resulting from the dense woolly hairs on the new leaves help to distinguish it.

UNDERGREEN WILLOW

Alaska Bog Willow
Salix fuscescens Anderss.

SYNONYM *Salix arbutifolia* Pall.

DESCRIPTION Trailing shrubs only 4–12 in. (10–30 cm) high. **LEAVES** typically obovate and rounded at tip, occasionally elliptic and pointed, 3/8–1½ in. (1–4 cm) long, margins entire or toothed near base, upper surface shiny dark green, lower surface whitish (glaucous). Petioles 1/8–1/4 in. (3–6 mm) long. **TWIGS** dark brown and smooth when young, becoming lighter with age.

CATKINS 3/4–1½ in. (2–4 cm) long, on leafy shoots developing with leaves, loose-flowered, dark purple; scales hairy, dark colored toward the tip.

SEED CAPSULES long and thin, on stalk 1/16–1/8 in. (1.5–3 mm) long, dark purple and with scattered hairs when young but becoming brown and hairless with age.

FLOWERING in June (at the same time as the leaves develop), fruits ripening in July.

HABITAT Alaska bog willow occurs commonly in wet tundra and small bogs beyond tree-line, and in moist, open black spruce stands, sedge meadows, and bogs throughout most of the Alaskan boreal forest; often growing with *S. barclayi, S. commutata, S. pulchra* and *S. myrtillifolia.*

TIP No other low-growing willow has the combination of smooth, shiny leaves broadest near the tip; and large, pear-shaped, red-hairy ovaries.

ALASKA BOG WILLOW

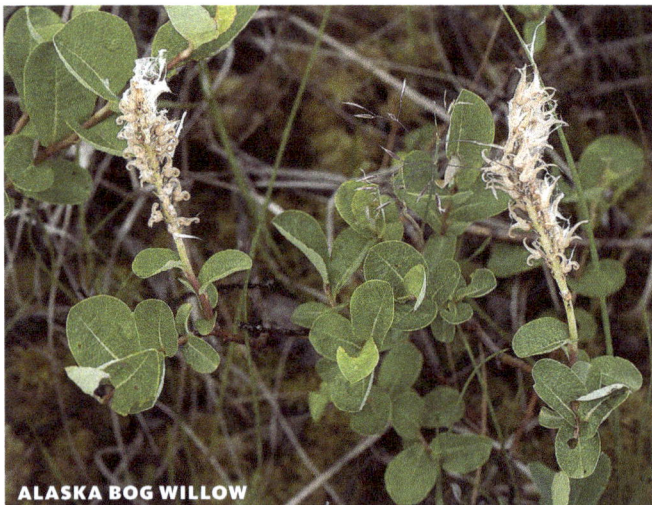

ALASKA BOG WILLOW

Grayleaf Willow
Salix glauca L.

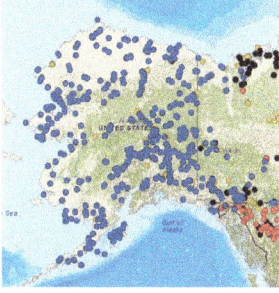

SYNONYMS *Salix cordifolia* Pursh

DESCRIPTION An erect to spreading shrub with a dull gray appearance, commonly 3–4 feet (1–1.2 m) tall but in exposed sites may be depressed; in favorable sites may become a small tree to 20 feet (6 m) high and 5 in. (12.5 cm) in trunk diameter. **LEAVES** variable in size, shape, and hairiness, oval to lanceolate, 1½–3½ in. (4–9 cm) long and 3/8–1½ in. (1–4 cm) wide, short-pointed to rounded at tip, margins usually entire but occasionally with small glandular teeth on the lower part. Upper surface green, densely hairy to nearly hairless; lower surface whitish (glaucous) with scattered hairs. Petiole 1/8–3/8 in. (3–10 mm) long. **Stipules** minute, glandular-margined, 1/32–5/16 in. (1–8 mm) long.

TWIGS reddish brown to grayish, hairy or hairless, with dense white hairs when young. **Winter buds** reddish brown with scattered hairs.

BARK gray, smooth, becoming rough and furrowed.

CATKINS 3/4–2 in. (2–5 cm) long, on leafy shoots, usually several inches back from end of branches developing with the leaves, persistent during most of the summer and often after leaves have fallen; scales light brown to yellow, rounded on tip, hairy on both surfaces.

SEED CAPSULE hairy, gray when young and turning light brown with age; on short stalk 1/32–1/16 in. (1–1.5 mm) long.

FLOWERING in June, fruits ripe in July and August. *Salix glauca* is generally a late-flowering species. In the subalpine zone, depending on local snow conditions, the catkins appear at the same time as the leaves. The mature female catkins may over-winter on the shrub, and the seeds released the following spring.

HABITAT Common after fire and as a pioneer species along rivers and roads and on glacial outwash, mine tailings, and abandoned fields in interior Alaska, in thickets with other willows. It also occurs as an individual open shrub in most forest types in the boreal forest. In the Arctic and western parts of Alaska, it grows on the floodplain and cutbanks of rivers, as well as protected sites in tundra habitats. Throughout Alaska except Aleutian Islands and southeast coast.

USES Because it is such a common species and seldom grows too tall to be reached by moose, grayleaf willow is an important browse species.

NOTE Though generally shrubby, grayleaf willow in Alaska reaches the size of a small, clump-forming tree. Hybrids with *Salix arctica* and *S. niphoclada* have been reported.

GRAYLEAF WILLOW

TIP Not always easily distinguished from *Salix niphoclada,* but *Salix glauca* leaves are wider and pointed at the tip, with longer petioles, stipules, and stipes. *Salix glauca* typically occurs in subalpine and alpine habitats; *Salix niphoclada* is most common in coastal wetlands.

GRAYLEAF WILLOW

GRAYLEAF WILLOW

Halberd Willow
Salix hastata L.

OTHER NAME Farr willow
SYNONYMS *Salix walpolei* (Coville & Ball) Ball, *Salix farrae* Ball var. *walpolei* Coville & Ball
DESCRIPTION A much branched spreading shrub 3–6 feet (1–2 m) high.
LEAVES elliptic, lanceolate to oblanceolate, 1–2 in. (2.5–5 cm) long and about 1/3 as wide, short-pointed, hairless, edges entire or with shallow teeth, upper surface yellow green to green, lower surface whitish (glaucous); reddish hairs usually present on upper surface midrib. Petiole slender, 1/16–5/16 in. (1.5–8 mm) long.
TWIGS reddish brown to brown, shiny, with dense white hairs when young.
CATKINS 3/4–1½ in. (2–4 cm) long, on leafy stalks, usually scattered along twig 2–4 in. (5–10 cm) from end, with the leaves; scales yellow at base and brown at tip, hairless or thinly hairy.
SEED CAPSULES 1/8–1/4 in. (3–6 mm) long on short stalks, light brown to reddish brown when mature.
FLOWERING in late June, seeds dispersed during late July and early August. Catkins develop at the same time as leaves.
HABITAT Occasional in the boreal forest of interior Alaska, primarily in willow thickets along small streams, but also as a pioneer species on river sandbars and on glacial moraines. It also occurs occasionally in alpine sedge bogs but does not seem to be abundant anywhere.
TIP Resembles *Salix barclayi*, but differs by the presence of scattered reddish hairs, especially on the leaf midrib.

HALBERD WILLOW

Coastal willow

Salix hookeriana Barratt ex Hook.

OTHER NAMES Hooker willow, bigleaf willow, Yakutat willow

SYNONYMS *Salix amplifolia* Cov.

DESCRIPTION A shrub or small tree, usually about 10–16 feet (3–5 m) tall but occasionally attaining a height of 25 feet (7.5 m) and a trunk diameter of 8–15 in. (20–38 cm).

LEAVES oval to broadly obovate or rarely the uppermost ovate, 1½–3 in. (4–7.5 cm) long and 3/4–2 in. (2–5 cm) wide, broadly pointed to rounded at tip, mostly rounded at base, edges without teeth or sparsely wavy-toothed, pale green above, whitish beneath, more or less hairy on both sides while unfolding but becoming hairless.

TWIGS stoutish, dark brown, densely white-or gray-woolly for 2 or 3 years. **Buds** dark reddish brown, hairy.

BARK gray, smooth.

CATKINS on leafy stalks, appearing before or with the leaves, 3–4 in. (7.5–10 cm) long and 1/2–5/8 in. (12–15 mm) wide at maturity; scales brownish to blackish, covered with long whitish hairs.

SEED CAPSULES long, hairless.

FLOWERING in mid-May to early June, seeds ripening mid-June to July.

HABITAT In Alaska, Hooker willow grows in a variety of sites including beach ridges, stabilized sand dunes, and coastal meadows. Rare in Alaska, except in the Yakutat Bay region where it was been known for many years as a local species, Yakutat willow (*Salix amplifolia* Cov.).

USES At Yakutat the plants are browsed by moose.

ETYMOLOGY This species honors William Jackson Hooker (1785–1865), English botanist, in whose work the description was published.

COASTAL WILLOW

Sandbar Willow
Salix interior Rowlee

SYNONYMS *Salix exigua* subsp. *interior* (Rowlee) Cronq.

DESCRIPTION An upright shrub, 10–12 feet (3–3.5 m) tall in Alaska, but becoming a small tree 20 feet (6 m) high in contiguous United States.

LEAVES long and very narrow, 1½–4 in. (4–10 cm) long, usually 1/4 in. (6 mm) wide, light green on both surfaces, edge sometimes entire but usually with sharp rather widely spaced teeth; petiole short.

TWIGS long, thin, unbranched, brown, and smooth.

CATKINS 1–2½ in. (2.5–6 cm) long on leafy stalks, appearing with the leaves; scales long, pale yellow, with thin hairs, and dropping soon after the catkin opens.

SEED CAPSULE long and slender, to 3/8 in. (10 mm) long on a short stalk.

FLOWERING in June, seeds ripening in late June and July.

HABITAT Occasional pioneer on the sand and silt bars of the rivers of interior Alaska, where it is often the first willow to invade a newly exposed bar, primarily by the development of shoots from its widely divergent root system. It seems to be unable to compete with other shrubs and trees, for it is seldom found in the older successional stages along the river and seldom reaches a height of more than 6–8 feet (2–2.5 m) in these localities.

USES Utilized as browse by moose, which often winter on the young islands and sandbars of the Tanana and Yukon Rivers.

SANDBAR WILLOW

SANDBAR WILLOW

Pacific Willow
Salix lasiandra Benth.

OTHER NAMES western black willow, yellow willow

DESCRIPTION A tall shrub or small tree to 20 feet (6 m) high. Farther south in contiguous United States, it is a small tree 20–30 feet (6–9 m) tall but occasionally a larger tree 50–60 feet (15–18 m) tall with a trunk 2–3 feet (60-90 cm) in diameter. LEAVES lance-shaped, 2–5 in. (5–12.5 cm) long and 1/2–1 in. (12–25 mm) wide, long pointed, mostly rounded at the base, with edges finely toothed, shiny green above, glaucous and more or less hairy beneath.

TWIGS hairy when young, stoutish, chestnut to reddish, shiny, hairless with age.

BARK gray, smooth, becoming rough and deeply furrowed.

WOOD pale brown, brittle.

CATKINS on leafy stalks, appearing with the leaves, 2–4 in. long (5–10 cm) at maturity; scales yellowish, hairy toward the base.Leaves and catkins develop together beginning in mid-May, maturing by the end of June.

SEED CAPSULES without hairs.

HABITAT Occasional pioneer species on the sand and silt bars of the rivers of interior Alaska, usually with other willows but occasionally forming pure stands. It is occasional to rare in the uplands in willow thickets along streams and roadsides.

USES Pacific willow is a preferred moose food. Useful for revegetation projects by planting dormant cuttings.

TIP The only willow species in Alaska having lance-shaped leaves and 4 or 5 stamens in each male flower.

PACIFIC WILLOW

Low Blueberry Willow
Salix myrtillifolia Anderss.

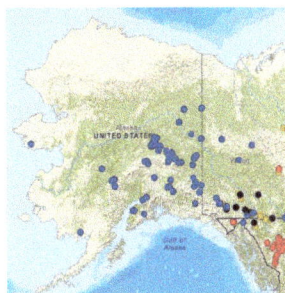

SYNONYM *Salix pseudocordata* (Anderss.) Rydb.

DESCRIPTION Low shrubs usually 8–24 in. (20–61 cm) tall, much branched and often prostrate.

LEAVES elliptic, ovate, to obovate, 3/4–1½ in. (2–4 cm) long and 3/8–3/4 in. (1–2 cm) wide, blunt or short-pointed at tip, margins toothed, upper surface dark green and shiny, lower surface slightly lighter, conspicuously net-veined. Petioles short; **stipules** small and inconspicuous and soon shedding.

TWIGS brown to gray, hairless, finely hairy when young.

CATKINS usually 3/4–1¼ in. (2–3 cm) long but occasionally to 2 in. (5 cm) long on leafy stalks appearing after the leaves have started to develop; scales brown to gray with long gray hairs.

SEED CAPSULES green to brown, hairless, 1/4–5/16 in. (6–8 mm) long on stalks 1/16–1/8 in. (1.5–3 mm) long.

HABITAT Occasional in black spruce stands and bogs in the interior of Alaska, often with dwarf birch (*Betula nana*). Locally abundant as a successional species following burning of low-lying black spruce stands. It also occurs occasionally in bogs below and just above tree-line.

TIP Distinguished from most shrubby willows by the green leaf underside. Leaf undersides of *Salix commutata* are also green but its leaves are hairy on both surfaces (especially the young leaves). Plants of *S. commutata* are typically taller, to 6 feet (2 m).

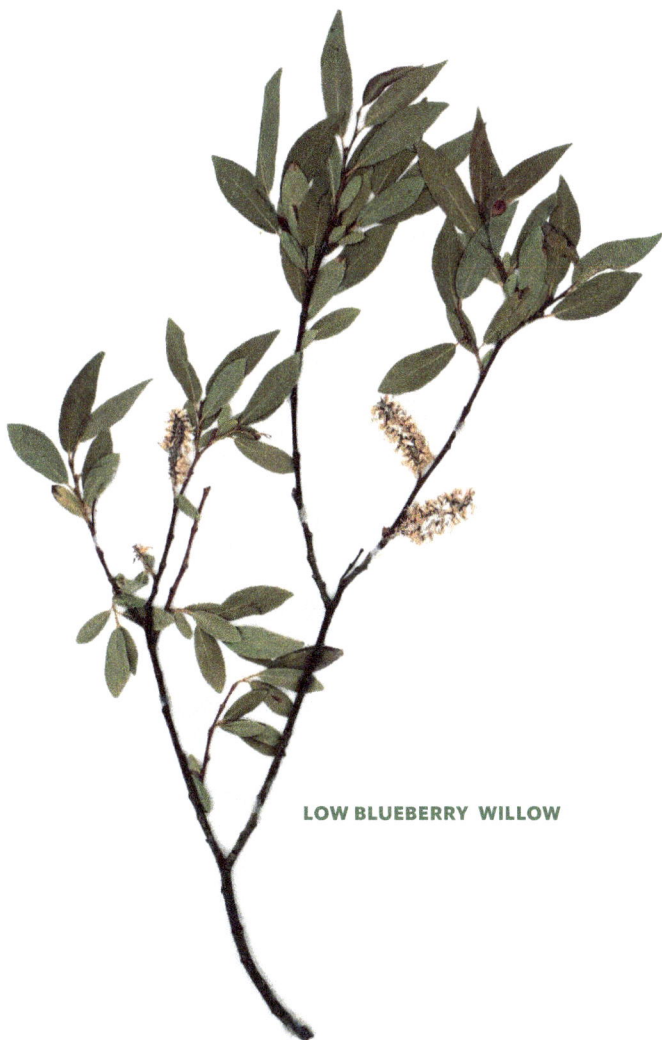

LOW BLUEBERRY WILLOW

Barren-Ground Willow
Salix niphoclada Rydb.

SYNONYM *Salix brachycarpa* Nutt. subsp. *niphoclada* (Rydb.) Argus

DESCRIPTION A low shrub with gray appearance, prostrate to erect, usually 1–3 feet (3–10 dm) tall but occasionally to 6 feet (2 m) in lowland and protected sites.

LEAVES variable, obovate to lanceolate, short-pointed, 1–1½ in. (25–40 mm) long, 3/16–3/8 in. (5–10 mm) wide, upper surface green, thinly hairy, lower surface whitish (glaucous), more thickly hairy. Petioles 1/32–1/8 in. (1–3 mm) long, reddish to yellow. **Stipules** glandular along edge, 1/16–1/8 in. (1.5–3 mm) long.

TWIGS thin, reddish brown to yellowish brown, densely hairy when young, becoming hairless. **Winter buds** reddish brown, hairy.

CATKINS 1–2½ in. (25–60 mm) long, on ends of leafy twigs, narrowly cylindrical, appearing gray from the dense hairs on flowers and scales; scales yellowish to dark brown, rounded, hairy on both surfaces.

SEED CAPSULE grayish green when young but becoming brown with age, thinly hairy on short stalk 1/32 in. (1 mm) long.

FLOWERING in June and July, seeds dispersed in July and August. Catkins developing with the leaves, and persisting throughout the summer and often through the following winter.

HABITAT Most common in arctic and alpine areas as a low shrub on talus slopes, in upland mountain-avens (*Dryas*) tundra, and in moist meadows and along stream margins. It occurs as a taller shrub, usually with other willows in the boreal forest, as a pioneer on well-drained alluvium and glacial outwash and moraine. *Salix niphoclada* is especially common in early successional stages along the Tanana River.

TIP Resembles *Salix glauca,* a willow of mostly higher elevations, and which has oval, pointed leaves, longer styles, and longer petioles to 5/8 in. (15 mm) long.

BARREN-GROUND WILLOW

BARREN-GROUND WILLOW

Ovalleaf Willow

Salix ovalifolia Trautv.

OTHER NAMES sprouting willow

SYNONYMS *Salix arctolitoralis* Hultén, *Salix flagellaris* Hultén, *Salix cyclophylla* Rydb.

DESCRIPTION Dwarf shrub only 3/4–2 in. (2–5 cm) tall; stems long and trailing.

LEAVES obovate to nearly round, 3/8–1 in. (1–2.5 cm) long and 1/2–4/5 as wide, on grooved petioles, upper surface green and glossy, lower surface pale green to whitish (glaucous), margins entire or slightly turned under, sometimes ciliate.

TWIGS slender, yellow to dark reddish brown, glabrous or sparsely pubescent toward tip, not glaucous, creeping and rooting at nodes.

CATKINS flowering as leaves emerge, 3/4–1½ in. (2–4 cm) long, on leafy shoots with scattered hairs when young, soon becoming hairless; scales reddish brown, hairy.

SEED CAPSULE smooth and greenish, whitish (glaucous) or reddish purple, becoming brown with age.

HABITAT Primarily in tundra along the Arctic, western and southwestern coasts. In the Arctic, most commonly along beaches and in saline meadows, more rarely in wet sites along rivers on inland sites, although never far from the sea.

NOTE Several varieties have been proposed, including var. *arctolitoralis* (Hultén) Argus, differing from var. *ovalifolia* in its overall larger size, with leaves narrowly elliptic, 1–2 in. (2.5–5 cm) long.

TIP Similar to creeping willow (*Salix stolonifera*), see key to distinguish the two species.

OVALLEAF WILLOW

OVALLEAF WILLOW

Skeletonleaf Willow
Salix phlebophylla Anderss.

DESCRIPTION A tiny, densely matted, prostrate shrub 1/2–1½ in. (1–4 cm) tall, usually with a thick central taproot.
LEAVES oblanceolate to obovate, 5/16–3/4 in. (8–20 mm) long and 3/16–3/8 in. (5–10 mm) wide, entire on edges, shiny green on both surfaces, with scattered hairs when young but becoming hairless with age. Veins 3–5 pairs from midrib and rather prominent beneath. New leaves crowded near tips of twigs, old leaves persisting at base one or more years, brown and partially skeletonized.
TWIGS radiating from central taproot, much branched and rooting sparingly.
CATKINS erect, 1/2–1 in. (12–25 mm) long, on short leafy shoots, developing with the leaves; scales broad, blunt, and black at tip, dark red at base, with long silky hairs.
SEED CAPSULES 1/4 in. (6 mm) long, on short stalk, densely silvery hairy when young but becoming nearly hairless at maturity.
HABITAT Common on dry usually exposed sites in the arctic and alpine tundra. It forms small dense mats along with a large number of mat-forming plants, especially mountain-avens (*Dryas*).
NOTE Closely related to least willow (*Salix rotundifolia*) but differs in having large, slightly hairy capsules, abundant skeletonized leaves, and 3–5 pairs of veins on leaves instead of 2–3 pairs in least willow. Numerous specimens appear to be intermediate between the two species.

SKELETONLEAF WILLOW

SKELETONLEAF WILLOW

Polar Willow
Salix polaris Wahlenb.

SYNONYM *Salix pseudopolaris* Flod.

DESCRIPTION A small prostrate shrub forming loose to some-times dense mats, 1–2 in. tall (2.5–5 cm), with branches often buried in moss or soil.

LEAVES oval to rounded, 1/4–3/4 in. (6–20 mm) long and 2/3 as wide, on short petioles; bright green on both surfaces to slightly paler beneath, margins entire, sparsely hairy beneath when young but becoming hairless with age.

TWIGS slender, rooted at nodes, often buried, smooth and red-dish.

CATKINS erect on leafy stalk 3/4–1½ in. (2–4 cm) long, developing with the leaves; scales broad and rounded at tip, brown to black and slightly lighter at base, with dense long hairs.

SEED CAPSULE broad, flask-shaped, 1/4 in. (6 mm) long, on short stalks, reddish-purple.

HABITAT Common to abundant in the arctic and alpine tundra of central and western Alaska. It forms loose mats in snow beds and along small streams where it is often imbedded in the moss and sedge mats.

NOTE Polar willow characteristi-cally forms a much looser mat than least willow (*Salix rotundifo-lia*) and skeletonleaf willow (*S. phlebo-phylla*), and is usually found on wetter sites.

POLAR WILLOW

POLAR WILLOW

Park Willow
Salix pseudomonticola C.R. Ball

OTHER NAMES cherry willow, serviceberry willow
SYNONYMS *Salix monticola* Bebb, *Salix padophylla* Rydb.,
DESCRIPTION An erect shrub, 3–12 feet (1–3.5 m) tall in Alaska but becoming a small tree southward in western Canada and northwestern United States.
LEAVES oval to elliptic, 1–4 in. (2.5–10 cm) long, 1/2–3/5 as broad, usually abruptly pointed to rounded at tip, glandular toothed on margins, purple to reddish yellow when young but soon turning green, shiny green and hairless above, whitish (glaucous) beneath, with prominent veins. Petioles 1/4–3/8 in. (6–10 mm) long. **Stipules** small and inconspicuous, or on fast growing shoots larger and leaflike with glandular teeth.

TWIGS yellow to reddish brown, shiny, hairy when young but becoming hairless.
BARK gray, smooth.
CATKINS short, 1¼–2½ in. (3–6 cm) long, stalkless twigs, appearing in May and early June before the leaves and usually shedding by end of June; scales 1/16 in. (1.5 mm) long, brown with long hairs.
SEED CAPSULES short and stout, 1/8–3/16 in. (3–5 mm) long, hairless, short-stalked, seeds ripening in June.
HABITAT A common pioneer willow on the braided rivers of interior Alaska and along other rivers and lake shores, forming thickets with other willows. Occasional in floodplain balsam poplar and spruce stands and in upland black spruce.
USES Along rivers it is utilized as a browse species by snowshoe hares and moose.
NOTE In early summer the reddish color of the new leaves stand out from the other willows.

PARK WILLOW

Tea-Leaf Willow
Salix pulchra Cham.

SYNONYM *Salix planifolia* Pursh subsp. *pulchra* (Cham.) Argus
DESCRIPTION An upright much branched shrub 3–6 feet (1–2 m) tall, rarely to 15 feet (4.5 m), often forming loose thickets in wet habitats but becoming a prostrate creeping shrub in exposed sites in arctic and alpine tundra.
LEAVES elliptic to oblanceolate, pointed at both ends and often diamond-shaped, as stated in the name, 1–2½ in. (2.5–6 cm) long and about 1/3–1/2 as wide, hairless and shiny green above and pale to whitish (glaucous) beneath, entire on edges or with a few small teeth near base. Petioles 1/8–3/8 in. (4–10 mm) long, slender. **Stipules** 1/4–1/2 in. (6–12 mm) long, linear, glandular-toothed, persisting on twigs 2–3 years. A few brown leaves usually remain on the twigs through the following winter and into the next summer.

TWIGS shiny dark brown, reddish or purple, hairy when young but becoming hairless in age.
BARK dark gray, smooth.
CATKINS 3/4–1½ in. (2–4 cm) long stalkless on the branches, developing in early spring before the leaves; scales blackish in upper part and hairy.
SEED CAPSULES 5/16 in. (8 mm) long, stout, hairy, greenish gray when young but becoming brown with age, on a short stalk.
FLOWERING in late May and early June, seeds ripening in late June and July; catkins shedding by August.
HABITAT Common shrub in bogs and other wet sites in the boreal forest of Alaska, forming thickets usually 3–5 feet (1–1.5 m) tall. It is a more upright, often isolated shrub in black spruce stands. It also occurs in the arctic and alpine treeless regions along streams and in the tundra where it may become a prostrate shrub. Almost all of Alaska except the western Aleutians and the coastal forests of southeastern Alaska.
USES Alaska Natives eat the young leaves as a green, both raw and cooked. The leaves must be picked when young, or they have a bitter

TEA-LEAF WILLOW

taste. In winter the twigs are browsed by moose and snowshoe hare, and the persistent leaves are often eaten by Dall sheep.
NOTE Diamondleaf willow is one of the few willows that can usually be identified in the winter condition; the shiny red twigs, the persistent stipules, and the persistent brown leaves are characteristic.
ADDITIONAL SPECIES The closely related **planeleaf willow** (*Salix planifolia* Pursh subsp. *planifolia*) has been reported from northern and southeast Alaska (see first key).

TEALEAF WILLOW

Netleaf Willow
Salix reticulata L.

OTHER NAMES thickleaf willow, reticulate willow

DESCRIPTION Prostrate creeping shrub, rooting along branches and ascending only a few inches; not a dense mat former.

LEAVES nearly round to oval, to 1½ in. (4 cm) long, thick and leathery, prominently net-veined on both surfaces but more conspicuously beneath; margins entire, upper surface green and roughened, lower surface whitish (glaucous) with scattered hairs along veins, petioles slender, red, 1/4–3/4 in. (6–20 mm) long.

TWIGS coarse, purplish when young, becoming reddish brown with age.

CATKINS erect, long and slender, to 2 in. (5 cm) long, on long leafless stalks; scales rounded, with long hairs on inner surface and nearly hairless on outer.

SEED CAPSULES stout, reddish, with short white hairs.

FLOWERING in June, seeds dispersing July to August.

HABITAT Common in a wide variety of vegetation types throughout most of Alaska, although it is more common in tundra than in forests. It grows on both dry and wet sites in the arctic and alpine tundra. In the boreal forest, it is most common around the edges of bogs and on hum-

NETLEAF WILLOW

mocks within the bogs, but it sometimes grows in open stands of black and white spruce, usually near timberline.

NOTE Netleaf willow is easily recognized by its thick round leaves with the net-veined pattern and by its long, slender, reddish catkins.

NETLEAF WILLOW

Richardson's willow
Salix richardsonii Hook.

OTHER NAMES woolly willow

SYNONYM *Salix lanata* L. subsp. *richardsonii* (Hook.) A. Skwortz.

DESCRIPTION Erect much-branched shrubs usually forming dense clumps 3–6 feet (1–2 m) tall, sometimes to 15 feet (4.5 m).

LEAVES elliptic to obovate, 3/4–2½ in. (2–6 cm) long, about 1/2 to 3/4 as wide, short-pointed or rounded at tip, entire or toothed on margins, both surfaces with long thin hairs when young but becoming hairless with age, dark green above, whitish (glaucous) beneath. Petioles stout, 1/8–3/8 in. (3–10 mm) long. **Stipules** conspicuous, long and narrow, with glandular teeth on the edges, persistent on the twig for several years.

YOUNG TWIGS stout and densely hairy, dark brown to black; older twigs hairless, orange-red to red-brown and characterized by persistent stipules.

BARK gray, smooth.

CATKINS 1½–2½ in. (4–6 cm) long on leafless peduncles, developing early in spring before the leaves; scales dark brown to black with dense silky hairs.

SEED CAPSULES stout, green to light brown, hairless, on short stalks.

FLOWERING in May and early June, seeds ripening in July, catkins shedding by August.

HABITAT A common thicket-forming shrub of stream banks and moist slopes in the Arctic and above timberline where it is often associated with alders and shrub birch, also in open spruce stands and old burns at lower elevations. In southeastern Alaska only in mountains in area from Juneau to Haines.

RICHARDSON'S WILLOW

RICHARDSON'S WILLOW

Least Willow
Salix rotundifolia Trautv.

DESCRIPTION A densely matted prostrate shrub, usually from a central taproot, forming mats about 1 in. (2.5 cm) high.

LEAVES ovate, less than 1/2 in. (12 mm) long, and 3/4 to 4/5 as wide, entire, shiny green on both surfaces, 2–3 pairs of veins prominent beneath; some dry leaves persistent about 1 year.

TWIGS radiating from a central taproot, slender, much branched, and rooting at nodes.

CATKINS short and few-flowered, 1/2 in. (12 mm) long, with 6–10 flowers; scales dark brown to black, thinly hairy.

SEED CAPSULES hairless, to 1/4 in. (6 mm) long.

HABITAT Arctic and alpine tundra of much of Alaska, but uncommon in the southeast.

LEAST WILLOW

LEAST WILLOW

Scouler Willow

Salix scouleriana Barratt ex Hook.

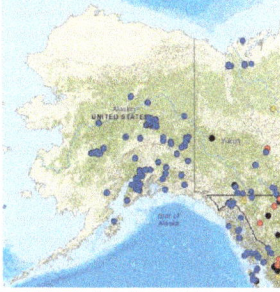

OTHER NAMES mountain willow, black willow, fire willow

DESCRIPTION A shrub or tree with compact rounded crown usually 15 feet (4.5 m) tall and 4 in. (10 cm) in trunk diameter but in some localities in Alaska becoming a tree 50–60 feet (15–18 m) tall and 16–20 in. (40.5–51 cm) in trunk diameter.

LEAVES variable, mostly oblanceolate to narrowly obovate or sometimes oblong or elliptic, 2–5 in. (5–12.5 cm) long and 1/2–1½ in. wide (12–40 mm), mostly very short-pointed at tip and tapering to base, edges without teeth to sparsely wavy-toothed, dark green and nearly hairless above, beneath whitish to white and more or less gray hairy or becoming rusty hairy when older.

TWIGS stoutish, yellowish or greenish brown and densely hairy when young, reddish to dark brown and nearly hairless when older; **buds** red.

BARK gray smooth, thin, becoming dark brown, divided into broad flat ridges.

WOOD light brown tinged with red and with thick whitish sapwood, fine-textured, lightweight, soft.

CATKINS stout, stalkless or on short leafless stalks, appearing in great abundance before the leaves, at maturity 1–2 in. (2.5–5 cm) long and nearly 1/2 in. (12 mm) thick; scales obovate, black, long hairy.

SEED CAPSULES long, slender, gray-woolly. One of the earliest flowering of the willows, its catkins developing as pussy willows even before the snow has melted.

FLOWERING in May, seeds dispersing in June, catkins shedding by July.

HABITAT Scouler willow is the most common willow of southeastern and south central Alaska where it occurs over a wide range of habitats and vegetation types. It is especially abundant in the vicinity of Anchorage and Kenai Peninsula where it has become widespread in the uplands following past widespread fires. It is often called "fire willow" because of its rapid occupation of burned areas, forming blue-green thickets. In southeastern Alaska it comes in abundantly after logging and also occurs along streams and roadsides and occasionally in the more open spruce and hemlock stands. Over all of southeastern and south central Alaska, it commonly reaches tree size. In the interior of Alaska, Scouler willow occurs in spruce, birch, and aspen stands, and occasionally in bogs, but is most common in areas that have been burned.

SCOULER WILLOW

USES In south-central Alaska it is an important moose browse species, and most trees have been barked by moose. One of several willows used for "diamond willow" carvings.

ETYMOLOGY This species honors its discoverer, John Scouler (1804–71), Scotch naturalist who made plant collections on the northwest coast of North America in 1825–27.

COULER WILLOW

Setchell Willow
Salix setchelliana Ball

DESCRIPTION A semi-prostrate, loose shrub with branches sometimes ascending to 12 in. (30 cm) long.

LEAVES obovate or oblanceolate, 1–2½ in. (2.5–6 cm) long, 3/8–3/4 in. (1–2 cm) wide, thick and fleshy, tapering to a petiole 1/8–1/4 in. (3–6 mm) long, rounded at tip, margins entire or irregularly glandular-toothed. Upper surface greenish blue and shiny, lower surface pale green to whitish (glaucous), with long silky hairs when young but becoming hairless.

TWIGS waxy, with dense long hairs when young but becoming smooth and gray, coarse.

CATKINS thick and fleshy, ¾ –1¼ in. (2–3 cm) long and 1/2 in. (12 mm) thick; scales large and conspicuous greenish yellow, with hairs on margins.

SEED CAPSULES thick and large, to 3/8 in. (10 mm) long, greenish yellow, turning brown with age.

HABITAT Almost totally restricted to the gravel outwash of the glacial rivers of the Alaska Range and adjacent mountain ranges.

NOTE Setchell willow is unique in Alaska because of its thick fleshy leaves and catkins. It does not seem to be closely related to any other willow.

ETYMOLOGY Named for William Albert Setchell (1864–1943), California botanist who made a collection of Alaska willows.

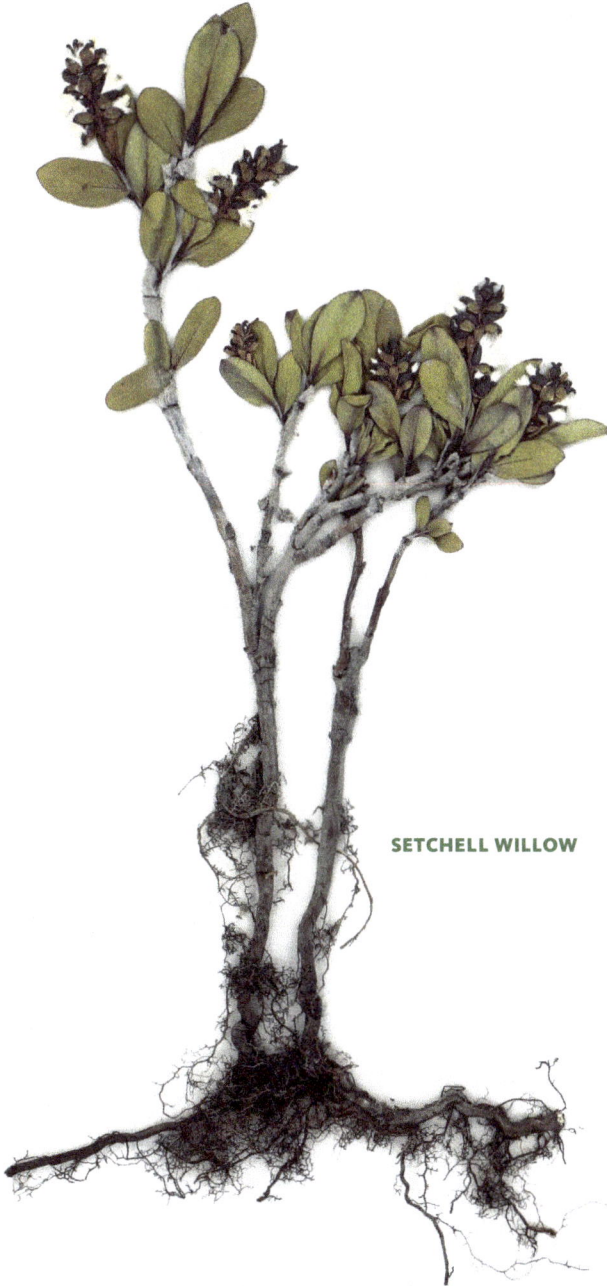

SETCHELL WILLOW

Sitka Willow
Salix sitchensis Sanson ex Bong.

OTHER NAMES silky willow

SYNONYM *Salix coulteri* Anderss.

DESCRIPTION A large shrub or small tree 10–20 feet (3–6 m) high with trunk 4–6 in. (10–15 cm) in diameter or rarely 30 feet (9 m) tall and 12 in. (30 cm) in diameter. In exposed places, becoming a low, nearly prostrate shrub.

LEAVES oblanceolate or narrowly obovate or sometimes elliptic, 2–4 in. (5–10 cm) long, usually short-pointed at tip, mostly tapering to a narrow base, edges without teeth or sparsely and inconspicuously wavy-toothed, above dark green and with sparse short hairs when young, beneath paler and with short silvery, silky hairs.

TWIGS slender, sometimes thinly hairy when young but when older hairless and dark reddish brown.

BARK gray, smooth, becoming slightly furrowed and scaly.

WOOD pale red, fine-textured, light-weight, soft.

CATKINS slender, tightly flowered on short leafy stalks, appearing with the leaves, 2–4 in. (5–10 cm) long at maturity; scales small, brown, densely hairy.

SEED CAPSULES short, silvery hairy.

FLOWERING in May, seeds ripening in early to mid-June, catkins shedding by July or early August.

HABITAT Common in the coastal forest region of southeast Alaska, growing in open locations along streams and beaches or in the upland where the forest is open or absent.

USES The wood is not used commercially though the Alaska Natives burn it in drying fish, as the smoke has no bad odor. The supple twigs have been used by Alaska Natives in basketmaking and for stretching skins, and the pounded bark has also been applied to heal wounds.

NOTE The satiny sheen on the lower surface of the leaves helps distinguish Sitka willow from other willows.

ETYMOLOGY Sitka willow was named for Sitka, Alaska, near which it was discovered by Karl Heinrich Mertens in 1827.

SITKA WILLOW

Creeping Willow
Salix stolonifera Cov.

DESCRIPTION Dwarf shrub, stems trailing or erect.
LEAVES obovate or elliptical, to 1¾ in. (4 cm) long, upper surface glabrous, glossy; undersurface sparsely hairy to glabrous; margins entire or irregularly glandular or glandular-toothed in the lower half.
TWIGS yellow- or greenish brown, not glaucous, glabrous and glossy.
CATKINS appearing with the leaves, to about 1¾ in. (4 cm) long, on stalks ½–2½ in. (1–6 cm) long; bracts obovate or rounded, brown, white-haired.
SEED CAPSULES glabrous, 1/8–3/8 in. (3–9 mm) long; style about 1/16 in. (1.5 mm) long.
HABITAT Arctic, subarctic, and alpine tundra; wet sedge meadows.
NOTE Similar to *Salix ovalifolia,* but differs in having long styles and thick branches; also similar to *S. fuscescens,* but differs in having long slender styles and entire leaf margins.
NOTE *Salix stolonifera* and *S. arctica* are reported to hybridize where found together.

CREEPING WILLOW

■ SANDALWOOD FAMILY *Santalaceae*

Parasitic dwarf shrubs on woody plants, with jointed brittle stems, brown, yellow, or green. **Leaves** opposite, small or reduced to scales. **Flowers** small, male and female on different plants (dioecious), calyx 2–6-parted, corolla none, stamens as many as parts of calyx, pistil with 1-celled inferior ovary and stigma. **Fruit** a berry, often sticky. Only one species in Alaska.

Western Hemlock Dwarf-Mistletoe
Arceuthobium tsugense (Rosendahl) G.N. Jones

OTHER NAMES dwarf-mistletoe

SYNONYM *Arceuthobium campylopodum* Engelm. f. *tsugense* (Rosend.) Gill.

DESCRIPTION Parasitic dwarf shrub on twigs, lower branches, and trunks of hemlock trees, greenish to reddish or brownish, usually inconspicuous, hairless. **Stems** slightly fleshy, of short thick angled joints enlarged at nodes, brittle. **Male plants** 1½ – 4 in. (4–10 cm) high, much branched; **female plants** smaller, less branched.

LEAVES reduced to paired brownish scales 1/16 in. (1.5 mm) long, joined at base in ring around twig.

FLOWERS minute, paired and stalkless or nearly so at sides of twig; **male flowers** less than 1/8 in. (3 mm) broad, yellowish, with 3–4 sepals and 3–4 stamens; **female flowers** about 1/16 in. (1.5 mm) broad, with 2 persistent sepals and pistil with inferior ovary and style.

FRUIT an elliptic flattened bluish berry 1/8–3/16 in. (3–5 mm) long on curved stalk, with mucilaginous or sticky flesh, discharging or shooting the sticky seed suddenly with force to about 20 feet (6 m) distance.

FLOWERING in August–September, fruit maturing the following September.

NOTE The deformed branches of infected trees, including witches-brooms (dense broomlike masses), swollen limbs, and swollen twigs, aid in recognition. However, these symptoms may have other causes. Also, there are cup scars after the limbs die back. Large burls or swellings are formed by trunk infections.

Western hemlock dwarf-mistletoe is Alaska's only parasitic woody plant. Western hemlock (*Tsuga heterophylla*) is the commonest host. However, this species occurs also on mountain hemlock (*Tsuga mertensiana*) and very rarely on Sitka spruce (*Picea sitchensis*). Southward it has been found on pines, firs, and other kinds of spruces, and a number of subspecies have been proposed.

This parasite is of considerable economic importance in southeast Alaska, though estimates of the damage are not available. Growth of hemlocks is slowed somewhat, but the trees are not killed. Many old-growth stands are infected, while others are not. Clearcutting infected stands is sometimes used to control the spread of this species. To remove the seed source of the parasite and to be effective, infected understory plants must be cut. Elsewhere, the slash is sometimes burned.

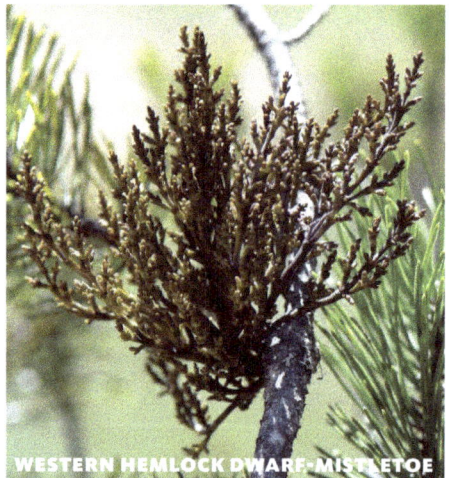

WESTERN HEMLOCK DWARF-MISTLETOE

■ SOAPBERRY FAMILY *Sapindaceae*

The Soapberry family is represented in Alaska by Douglas maple in the southeast. Maples, formerly placed in the Aceraceae but now merged into Sapindaceae, have the following characteristics: (1) **leaves** paired (opposite), long-stalked, broad, 3-lobed or 5-lobed and toothed; (2) **flowers** male and female on the same or different trees, small, in clusters appearing with the leaves; and (3) **fruits** distinctive, paired winged, 1–seeded keys. In winter the paired (opposite) U-shaped leaf-scars aid in recognition.

Douglas Maple
Acer glabrum Torr.

OTHER NAMES dwarf maple, Douglas Rocky Mountain maple
SYNONYM *Acer douglasii* Hook.
DESCRIPTION Small deciduous tree of southeast Alaska, becoming 20–30 feet (6–9 m) tall and 6–12 in. (15–30 cm) in trunk diameter but often a several-stemmed shrub 4–6 feet (1.2–1.8 m) high.
LEAVES paired (opposite), ovate, 2–4 in. (5–10 cm) long and broad, slightly heart-shaped at base, shallowly 3-lobed with the lobes long-pointed, deeply, sharply, and irregularly or doubly toothed, hairless, shiny dark green above, pale beneath with yellowish veins. Petioles 1½–4 in. (4–10 cm) long, slender, reddish tinged.
TWIGS paired (opposite), reddish, hairless, with U-shaped leaf scars. **Winter buds** short-pointed, 1/8–1/4 in. (3–6 mm) long, dark red, the side buds paired (opposite).
BARK gray, smooth.
WOOD light brown, heavy, hard, fine-grained.
FLOWER CLUSTERS (corymbs) terminal, appearing with the leaves, with several flowers on slender spreading or drooping stalks. **Flowers** mostly male and female on different trees (dioecious), about 1/8 in. (3 mm) long, composed of 4 narrow yellow green sepals as long as the narrow yellow green petals, 7–8 stamens, and in female flowers very short stamens and pistil with 2-celled ovary and 2 styles.
FRUIT of paired, winged, 1-seeded keys (samaras) 3/4–1 in. (2–2.5 cm) long, usually red until shed, then turning to light brown.
FLOWERING in May, fruit maturing July–August.
HABITAT Common along shores in southeast Alaska, sometimes fringing tidal meadows or bogs. Occasional in rich moist soils on forested slopes.
USES The trees are seldom large enough for commercial purposes.

DOUGLAS MAPLE

ETYMOLOGY Named for its discoverer, David Douglas (1798–1834), Scotch botanical explorer, who introduced many trees from western North America to Europe.

ADDITIONAL SPECIES **Vine Maple** (*Acer circinatum* Pursh) is reported to occur in southeast Alaska but its presence is not verified. **Bigleaf maple** (*Acer macrophyllum* Pursh) is known from southeastern British Columbia but has not yet been discovered in Alaska. Bigleaf maple is distinguished by the paired (opposite) long-stalked, very large leaves 5–14 in. (12.5–35.5 cm) long and broad, which are heart-shaped, deeply 5-lobed with additional smaller lobes, and with few teeth. The clustered fruits are paired, winged 1–seeded keys 11/4–2 in. (3–5 cm) long and 1/2 in. (1.2 cm) wide, bristly hairy at their base.

VINE MAPLE

DOUGLAS MAPLE

IDENTIFICATION KEYS

Keys are provided to aid identication, both in summer, when leaves, flowers, and fruits are present, and in winter, when twigs, winter buds, and bark are used. Included here are:
- **Key to Alaska trees based mainly onleaves,**
- **Winter key to deciduous trees of Alaska,**
- **Key to genera of Alaska shrubs,** and
- **Winter key to Alaska shrubs.**

Also, each genus with two or more species has a key to its Alaska species. For the willows (*Salix*), the largest genus, there are two: **Key to Alaska willows** and **Vegetative key to Alaska willows.**

A key is an outline for identifying specimens or plants through the process of elimination. This device is a short cut to save time in reading every description until the one that agrees is found. The species are divided into two groups according to certain distinguishing characters, and each group is divided successively into two smaller groups down to a single species at the end. The name of a particular specimen is found through selection, one by one, of the group in which it belongs. Then verification is made by comparison with description, illustrations, and maps. If agreement is lacking or doubtful, the steps followed in the key may be retraced and different steps tried.

Emphasis is given to nontechnical and vegetative characters, which are present over longer periods than flowers or fruits; also, to the larger plant parts. However, a hand lens will be helpful for observing details.

The first step is to select the proper key for the specimen, whether it is a tree or shrub and in summer or winter condition. Usually, the keys based mainly on leaves and other vegetative characters are simpler and easier to use than the winter keys. The latter must depend largely upon differences in twigs and buds. Even in winter though, enough old leaves, flowers, and fruits may be found to allow use of the main keys. Of course, keys based largely on leaves can be used throughout the year for the evergreens.

KEY TO ALASKA TREES BASED MAINLY ON LEAVES

1 Leaves needlelike or scalelike, evergreen (except in tamarack), trees resinous (except in yew); seeds more or less exposed and not enclosed in a fruit (conifers or softwoods, i.e., gymnosperms) **2**

1 Leaves broad and flat, shed in fall (deciduous); trees non-resinous; seeds developed from a flower and enclosed in a fruit flowering plants (angiosperms) **14**

CONIFERS

2 Leaves needlelike, flattened, abruptly pointed but not prickly, in 2 rows, comb-like with leafstalks extending down twig; seeds single in scarlet juicy cuplike disk; uncommon in southeast Alaska **PACIFIC YEW** (*Taxus brevifolia, p. 45*)

2 Leaves needlelike or scalelike, not as above; seeds borne on scales of a cone **3**

3 Leaves needlelike, more than 1/4 in. (6 mm) long, single, or clustered **4**

3 Leaves scalelike, usually less than 1/8 in. (3 mm) long, crowded, forming fanlike or flattened sprays **13**

4 Needles shedding in fall, 12-20 in cluster on short spur twigs (also single on leading twigs) **TAMARACK** (*Larix laricina, p. 28*)

4 Needles evergreen, single or 2 (sometimes 3) in a bundle **5**

5 Needles 2 (sometimes 3) in a bundle with sheath at base **LODGEPOLE PINE** (*Pinus contorta, p. 38*) **6**

5   Needles single, without sheath at base   7
6   Cones pointing backward, opening at maturity; generally low spreading tree of muskegs in coastal forests   SHORE PINE (*Pinus contorta* var. *contorta, p. 39*)
6   Cones pointing outward, mostly remaining closed many years; tree often tall and narrow of inner fiord forests at head of Lynn Canal (Skagway to Haines)
                              LODGEPOLE PINE (*Pinus contorta* var. *latifolia, p. 39*)
7   Older twigs roughened by projections where needles were shed   8
7   Older twigs smooth FIR (*Abies*)   12
8   Needles sharp-pointed, stiff, without leafstalks SPRUCE (*Picea*)   9
8   Needles blunt, soft and not stiff, with short leaf-stalks HEMLOCK (*Tsuga*)   11
9   Needles flattened but slightly keeled   SITKA SPRUCE (*Picea sitchensis, p. 35*)
9   Needles 4-angled   10
10 Twigs hairy; needles mostly less than 1/2 in. (12 mm) long, resinous; cones egg-shaped or nearly round, mostly less than 1 in. (2.5 cm) long, curved down on short stalks, remaining on tree   BLACK SPRUCE (*Picea mariana, p. 33*)
10 Twigs hairless; needles more than 1/2 in. (12 mm) long, with skunklike odor when crushed; cones cylindric, 1¼ –2½ in. (3–6 cm) long, falling at maturity
                                       WHITE SPRUCE (*Picea glauca, p. 30*)
11 Needles flat, appearing in 2 rows, shiny dark green above, with 2 whitish bands (stomata) on lower surface   WESTERN HEMLOCK (*Tsuga heterophylla, p. 41*)
11 Needles half-round and keeled or angled beneath, crowded on all sides of short side twigs, blue green, with whitish lines (stomata) on both surfaces
                              MOUNTAIN HEMLOCK (*Tsuga mertensiana, p. 43*)
12 Needles shiny dark green on upper surface and silvery white with many lines (stomata) on lower surface   PACIFIC SILVER FIR (*Abies amabilis, p. 24*)
12 Needles dull dark green with whitish lines (stomata) on both surfaces
                                   SUBALPINE FIR (*Abies lasiocarpa, p. 26*)
13 Leafy twigs flattened; leaves flattened and curved, not spreading
                              WESTERN REDCEDAR (*Thuja plicata, p. 21*)
13 Leafy twigs 4-angled or slightly flattened; leaves pointed, spreading
                              ALASKA-CEDAR (*Callitropsis nootkatensis, p. 17*)

BROADLEAF TREES

14 Leaves and twigs arranged singly (alternate)   15
14 Leaves and usually twigs in pairs (opposite)   36

Leaves and twigs alternate

15 Leaves not divided into leaflets (simple)   16
15 Leaves divided into 7–17 leaflets (compound), the leaflets attached along extended leafstalk and shedding with it MOUNTAIN-ASH (*Sorbus*)   34
16 Leafstalks (petioles) mostly less than 1/2 in. (12 mm) long; leaves mostly more than twice as long as broad, with edges finely toothed or without teeth; winter buds covered by a single scale WILLOW (*Salix*)   17
16 Leafstalks (petioles) mostly more than 1/2 in. (12 mm) long (often shorter in alder); leaves less than twice as long as broad, with edges finely or coarsely toothed; winter buds with 2 or more scales exposed   24
17 Leaf edges without teeth or only sparsely and indistinctly toothed   18
17 Leaf edges finely and distinctly toothed from base to tip   23
18 Leaves rounded at base, broadly elliptic, becoming hairless on both sides
                                   COASTAL WILLOW (*Salix hookeriana, p. 206*)
18 Leaves tapering or short-pointed at base, narrower, with hairs on lower surface   19

19 Lower surface of leaves covered by dense hairs, appearing silvery, white, or gray **20**
19 Lower surface of leaves visible through less dense hairs **22**
20 Leaves thick, lower surface with dense white woolly hairs
  FELTLEAF WILLOW (*Salix alaxensis, p. 187*)
20 Leaves thin, lower surface with dense straight hairs **21**
21 Lower surface of leaves silvery silky hairy, upper surface green with scattered hairs
  SITKA WILLOW (*Salix sitchensis, p. 224*)
21 Lower surface of leaves dull gray hairy, upper surface greenish gray and hairless
  GRAYLEAF WILLOW (*Salix glauca, p. 203*)
22 Leaves thick, nearly hairless above; hairs on lower surface short and stiff, at least some red, giving reddish hue **SCOULER WILLOW** (*Salix scouleriana, p. 221*)
22 Leaves thin, hairy on both sides; hairs longer, not reddish
  BEBB WILLOW (*Salix bebbiana, p. 196*)
23 Leaves 1–3 in. (2.5–7.5 cm) long, mostly short-pointed at both ends
  LITTLETREE WILLOW (*Salix arbusculoides, p. 188*)
23 Leaves 2–5 in. (5–12.5 cm) long, long-pointed, mostly rounded at base
  PACIFIC WILLOW (*Salix lasiandra, p. 208*)
24 Leaf edges finely toothed with curved and rounded teeth **COTTONWOOD, POPLAR, ASPEN** (*Populus*) **25**
24 Leaf edges coarsely toothed with sharp-pointed teeth **27**
25 Leaf blades nearly round, less than 2 in. (5 cm) long; leafstalks flattened
  QUAKING ASPEN (*Populus tremuloides, p. 176*)
25 Leaf blades longer than broad, 2½–5 in. (6–12.5 cm) long; leafstalks round **26**
26 Seed capsules pointed, hairless, 2-parted; leaves pale green and brownish beneath; tree of interior forests **BALSAM POPLAR** (*Populus balsamifera, p. 174*)
26 Seed capsules rounded, hairy, 3-parted; leaves whitish beneath; tree of coastal forests
  BLACK COTTONWOOD (*Populus trichocarpa, p. 178*)
27 Leaf edges doubly toothed with teeth of 2 sizes **28**
27 Leaf edges with uniform teeth **33**
28 Leaf edges not lobed; bark papery and peeling off, white, brown, or pinkish **BIRCH** (*Betula*) **29**
28 Leaf edges wavy or shallowly lobed; bark usually gray and smooth, not papery nor peeling off **ALDER** (*Alnus*) **31**
29 Leaves long-pointed, usually wedge-shaped at base; bark usually white in age; mostly interior Alaska **ALASKA PAPER BIRCH** (*Betula neoalaskana, p. 65*)
29 Leaves mostly short-pointed **30**
30 Leaves thin, mostly rounded at base; bark usually reddish brown
  PAPER BIRCH (*Betula papyrifera, p. 66*)
30 Leaves thick, wedge-shaped or rounded at base, with white hairs on toothed edges; bark usually dark brown or gray; southern and southern interior Alaska, absent from southeast Alaska **KENAI BIRCH** (*Betula kenaica, p. 63*)
31 Leaves yellow green above, shiny on both sides and especially beneath, sticky when young, edges with relatively long-pointed teeth **SITKA ALDER** (*Alnus viridis, p. 58*)
31 Leaves dark green above, dull, not sticky when young, edges with short-pointed teeth **32**
32 Leaves thick with edges curled under slightly, with rusty hairs along veins beneath
  RED ALDER (*Alnus rubra, p. 56*)
32 Leaves thin with edges flat, finely hairy or nearly hairless beneath
  THINLEAF ALDER (*Alnus incana, p. 55*)
33 Leaves short-pointed, sometimes 3-lobed **OREGON CRAB APPLE** (*Malus fusca, p. 151*)
33 Leaves rounded at tip **SASKATOON SERVICEBERRY** (*Amelanchier alnifolia, p. 138*)

34 Leaflets mostly 11–15, short-pointed, edges toothed nearly to base   **35**
34 Leaflets mostly 9 or 11, rounded or short-pointed at tip, edges not toothed in lowest third
                          SITKA MOUNTAIN-ASH (*Sorbus sitchensis, p. 168*)
35 Leaflets becoming hairless   CASCADE MOUNTAIN-ASH (*Sorbus scopulina, p. 167*)
35 Leaflets white-hairy beneath; naturalized tree
                          EUROPEAN MOUNTAIN-ASH (*Sorbus aucuparia, p. 164*)

Leaves and twigs opposite

36 Leaves with 3 long-pointed lobes, irregularly or doubly toothed
                          DOUGLAS MAPLE (*Acer glabrum, p. 227*)
36 Leaves divided into 5 or 7 leaflets (compound), finely toothed
                          PACIFIC RED ELDER (*Sambucus racemosa, p. 46*)

WINTER KEY TO ALASKAN DECIDUOUS TREES

1   Twigs with many wartlike, blackish spur twigs about 1/8 in. (3 mm) long; upright brown
    cones usually present; trees with pointed crown   TAMARACK (*Larix laricina, p. 28*)
1   Twigs without spur twigs or with longer spurs; trees with spreading, usually rounded
    crown   **2**
2   Winter buds, leaf-scars, and twigs arranged singly (alternate)   **3**
2   Winter buds, leaf-scars, and usually twigs in pairs (opposite)   **15**

Leaves and twigs alternate

3   Winter buds covered by a single scale willow (*Salix;* the species not readily distinguished
    in winter)
3   Winter buds with 2 or more scales exposed   **4**
4   Winter buds usually resinous or sticky, shiny, brown, long-pointed; lowest bud-scale cen-
    tered over leaf-scar COTTONWOOD, POPLAR, ASPEN (*Populus*)   **5**
4   Winter buds not resinous or sticky (slightly so in Sitka alder); lowest bud-scale at side of
    bud   **7**
5   Winter buds 1/4 in. (6 mm) or less in length, slightly or not resinous
                          QUAKING ASPEN (*Populus tremuloides, p. 176*)
5   Winter buds 3/8–1 in. (10–25 mm) long, very resinous   **6**
6   Tree of interior forests   BALSAM POPLAR (*Populus balsamifera, p. 174*)
6   Tree of coastal forests   BLACK COTTONWOOD (*Populus trichocarpa, p. 178*)
7   Winter buds mostly stalked (slightly so in *Alnus viridis*), with the 3 exposed scales meeting
    at edges (overlapping in *Alnus viridis*); old, hard, blackish cones or conelike fruits usually
    present ALDER (*Alnus*)   **8**
7   Winter buds not stalked, composed of overlapping scales; fruits not conelike   **10**
8   Cones with long stalks more than 1/2 in. (12 mm) long, mostly longer than cones
                          SITKA ALDER (*Alnus viridis, p. 58*)
8   Cones with short stalks less than 1/2 in. (12 mm) long   **9**
9   Cones 1/2–1 in. (12–25 mm) long   RED ALDER (*Alnus rubra, p. 56*)
9   Cones less than 1/2 in. (12 mm) long   THINLEAF ALDER (*Alnus incana, p. 55*)
10 Winter buds 1/4 in. (6 mm) or less in length; bud-scales with few or no hairs   **11**
10 Winter buds large, usually more than 3/8 in. (10 mm) long; inner exposed bud-scales
    densely hairy MOUNTAIN ASH (*Sorbus*)   **13**
11 Twigs with many small whitish dots (lenticels and resin); bark papery, peeling off
                          PAPER BIRCH (*Betula papyrifera, p. 66*)
11 Twigs with few inconspicuous dots (lenticels); bark not papery   **12**
12 Winter buds blunt-pointed, dark brown; twigs coarse, gray or brown, often with dense
    gray hairs near tip, with short side twigs or spurs
                          OREGON CRAB APPLE (*Malus fusca, p. 151*)

12 Winter buds sharp-pointed, purple; twigs slender, reddish purple, shiny, hairless, without short side twigs or spurs   SASKATOON SERVICEBERRY (*Amelanchier alnifolia, p. 138*)
13 Winter buds with whitish hairs   **14**
13 Winter buds with rusty brown hairs   SITKA MOUNTAIN-ASH (*Sorbus sitchensis, p. 168*)
14 Winter buds reddish brown, inner scales with whitish hairs at tip
CASCADE MOUNTAIN-ASH (*Sorbus scopulina, p. 167*)
14 Winter buds densely covered with whitish hairs; naturalized tree
EUROPEAN MOUNTAIN-ASH (*Sorbus aucuparia, p. 164*)

Leaves and twigs alternate

15 Twigs slender, reddish, with small dark red buds DOUGLAS MAPLE (*Acer glabrum, p. 227*)
15 Twigs stout, gray, with large gray buds   PACIFIC RED ELDER (*Sambucus racemosa, p. 46*)

KEY TO GENERA OF ALASKA SHRUBS

This summer key is for use with flowering specimens and is based upon some flower and fruit characters, as well as vegetative characters of twigs and leaves. Keys to species are included in the text for the genera with two or more species. For incomplete specimens, see the winter key to Alaska shrubs; it is more detailed, except for the willows (*Salix*), as they are very difficult to identify to the species-level without leaves and catkins.

1 Plants parasitic on conifer twigs; leaves reduced to paired brownish scales
HEMLOCK DWARF-MISTLETOE (*Arceuthobium tsugense, p. 226*)
1 Plants growing on land; leaves green   **2**
2 Shrubs without true flowers; seeds in persistent, berrylike, resinous, blue or green cones; leaves scalelike, awl-shaped, or needlelike, resinous   JUNIPER (*Juniperus, p. 19*)
2 Shrubs with flowers; seeds enclosed in fruits; leaves mostly broad (if needlelike or scalelike, fruit a capsule or black berry)   **3**
3 Flowers crowded in heads or catkins   **4**
3 Flowers not in heads or catkins   **8**
4 Flowers in dense yellow heads; leaves finely dissected, whitish hairy, with sagebrush odor
SAGEBRUSH (*Artemisia, p. 52*)
4 Flowers in catkins, long narrow clusters, male and female separate; leaves various   **5**
5 Fruit a capsule with many hairy seeds; bud covered by 1 scale   WILLOW (*Salix, p. 179*)
5 Fruit a nutlet, 1-seeded, not hairy, bud covered by 2 or more scales   **6**
6 Leaves aromatic, with minute resin dots, oblanceolate, rounded at tip and with several coarse teeth; male catkins erect   SWEETGALE (*Myrica gale, p. 136*)
6 Leaves not aromatic or resin dotted, elliptic or ovate, toothed along margin; male catkins drooping (BIRCH FAMILY, *Betulaceae*)   **7**
7 Leaves small, mostly less than 1 in. (2.5 cm) long, nearly as broad as long, twigs densely glandular   BIRCH (*Betula, p. 60*)
7 Leaves larger, mostly more than 2 inches (5 cm) long, longer than broad, pointed at tip; twigs not glandular   ALDER (*Alnus, p. 54*)
8 Leaves with minute scales; flowers with calyx but no corolla (elaeagnus family, Elaeagnaceae)   **9**
8 Leaves not scaly; flowers with both calyx and corolla   **10**
9 Leaves opposite, with brown scales   BUFFALOBERRY (*Shepherdia canadensis, p. 77*)
9 Leaves alternate, with silvery scales   SILVERBERRY (*Elaeagnus commutata, p. 76*)
10 Petals separate   **11**
10 Petals united, at least partly, into a corolla tube   **29**
11 Ovary or ovaries superior, with calyx and corolla attached below   **12**
11 Ovary inferior, with petals and sepals attached above; fruits fleshy   **22**

12 Ovaries few to many (rose family, Rosaceae)   13
12 Ovary 1   19
13 Fruits dry; stems without spines and prickles   14
13 Fruits fleshy; stems mostly with spines or prickles   18
14 Shrubs low or prostrate, less than 6 in. (15 cm) high   15
14 Shrubs upright, more than 12 in. (30 cm) high   16
15 Leaves twice divided into 3 narrow pointed segments, thin, hairless
   LUETKEA (*Luetkea pectinata, p. 150*)
15 Leaves oblong, leathery, densely white-hairy beneath   MOUNTAIN-AVENS (*Dryas, p. 143*)
16 Leaves pinnately compound; petals yellow BUSH CINQUEFOIL (*Dasiphora fruitcosa, p. 141*)
16 Leaves simple; petals white or pink   17
17 Leaves 3-5 lobed, palmately veined   PACIFIC NINEBARK (*Physocarpus capitatus, p. 152*)
17 Leaves not lobed, pinnately veined   SPIRAEA (*Spiraea, p. 170*)
18 Fruit a raspberry or similar, of crowded drupelets; leaves simple or divided into 3-5
leaflets   RASPBERRY, SALMONBERRY, THIMBLEBERRY (*Rubus, p. 158*)
18 Fruit a rose hip, fleshy and rounded enclosing the "seeds"; leaves pinnate with 5 or more
leaflets   ROSE (*Rosa, p. 154*)
19 Leaves less than 1/4 in. (6 mm) long, needlelike; fruit berrylike, blue black or purple
   CROWBERRY (*Empetrum nigrum, p. 94*)
19 Leaves larger and broader; fruit a capsule   20
20 Leaves thin, deciduous, with straight or entire margins
   COPPERBRUSH (*Elliottia pyroliflora, p. 93*)
20 Leaves thick and leathery, evergreen, with rolled or toothed margins   21
21 Leaves densely woolly beneath, rolled under on margins
   LABRADOR-TEA (*Rhododendron, p. 110*)
21 Leaves hairless beneath, sharply toothed on margins
   PIPSISSEWA (*Chimaphila umbellata, p. 92*)
22 Leaves opposite or paired   RED-OSIER DOGWOOD (*Cornus alba, p. 71*)
22 Leaves alternate or single   23
23 Low creeping shrubs; petals 4, red to pink, bent backward; fruit a cranberry
   BOG CRANBERRY (*Vaccinium oxycoccos, p. 120*)
23 Upright shrubs; petals 5, spreading   24
24 Leaves palmately veined and lobed   25
24 Leaves or leaflets pinnately veined; fruit like a small apple (pome) (rose family, Rosaceae)
   26
25 Leaves small, not prickly; twigs slender, mostly without spines or prickles
   CURRANT, GOOSEBERRY (*Ribes, p. 126*)
25 Leaves large, with prickles on veins; twigs stout, very spiny
   DEVIL'S-CLUB (*Oplopanax horridus, p. 50*)
26 Leaves pinnately compound with 7–17 leaflets   MOUNTAIN-ASH (*Sorbus, p. 164*)
26 Leaves simple   27
27 Leaves elliptic, rounded at tip, not lobed   SERVICEBERRY (*Amelanchier, p. 138*)
27 Leaves mostly ovate, pointed at tip, often lobed   28
28 Twigs usually bearing stout spines; fruit blackish
   BLACK HAWTHORN (*Crataegus douglasii, p. 140*)
28 Twigs sometimes ending in spines; fruit yellow or red
   OREGON CRAB APPLE (*Malus fusca, p. 151*)
29 Leaves alternate or single (heath family, Ericaceae)   30
29 Leaves opposite or paired   40
30 Fruit a berry or berrylike   31

30 Fruit a dry capsule   33

31 Ovary superior   32

31 Ovary inferior; berry blue or red; leaves entire or finely toothed on margin
BLUEBERRY, HUCKLEBERRY (*Vaccinium, p. 116*)

32 Fruit a berrylike capsule, covered by fleshy purplish or white calyx; leaves sharply or wavy-toothed on margin   SALAL, WINTERGREEN (*Gaultheria, p. 96*)

32 Fruit a drupe with 4–5 stony nutlets, red or blueblack; leaves not toothed on margin
BEARBERRY (*Arctostaphylos, p. 81*)

33 Shrubs more than 4 ft. (1.2 m) high; leaves thin, deciduous; twigs and leaves with glandular ("sticky") hairs   RUSTY MENZIESIA (*Menziesia ferruginea, p. 102*)

33 Shrubs less than 4 ft. (1.2 m) high; leaves thick, evergreen; twigs and leaves without glandular ("sticky") hairs   34

34 Upright shrubs, loosely branching, not forming mats; leaves not crowded, not needlelike, more than 1/2 in. (12 mm) long   35

34 Low shrubs forming dense mats; leaves crowded, needlelike   38

35 Leaves oblong to linear, edges rolled under   36

35 Leaves elliptic or oblanceolate, edges not or slightly rolled under   37

36 Corolla purple, saucer-shaped; leaves whitish beneath with inconspicuous veins
BOG KALMIA (*Kalmia polifolia, p. 99*)

36 Corolla pinkish to crimson, urn-shaped; leaves greenish or whitish beneath with conspicuous veins   BOG-ROSEMARY (*Andromeda polifolia, p. 79*)

37 Flowers erect or spreading, with saucer-shaped showy pink to purple corolla
ROSEBAY, RHODODENDRON (*Rhododendron, p. 110*)

37 Flowers hanging singly under twig, with bell-shaped white corolla
LEATHERLEAF (*Chamaedaphne calyculata, p. 90*)

38 Flowers usually several at stem tip, corolla yellow, blue, or red
MOUNTAIN-HEATH (*Phyllodoce, p. 104*)

38 Flowers usually single, corolla pink or white   39

39 Stems partly erect, to 6 in. (15 cm) high; flower stalk less than 1/2 in. (12 mm) high
STARRY CASSIOPE (*Cassiope stelleriana, p. 87*)

39 Stems creeping, forming dense mats less than 2 in. (5 cm) high; flower stalk ¾ –1½ in. (2–4 cm) long   DIAPENSIA (*Diapensia lapponica, p. 74*)

40 Ovary superior, with calyx and corolla attached below (heath family, Ericaceae)   41

40 Ovary inferior, with calyx and corolla attached above   43

41 Leaves scalelike, pressed against stems   CASSIOPE (*Cassiope, p. 85*)

41 Leaves larger, spreading   42

42 Stems creeping, forming dense mats to 2 in. (5 cm) high; leaves needlelike, 1/4 in. (6 mm) long   ALPINE-AZALEA (*Kalmia procumbens, p. 100*)

42 Stems upright, 4–20 in. (1–5 dm.) high; leaves oblong, ¾ –1½ in. (2–4 cm) long
BOG KALMIA (*Kalmia polifolia, p. 99*)

43 Stems creeping; leaves evergreen, rounded, 1/4–5/8 in. (6–15 mm) long and wide
TWINFLOWER (*Linnaea borealis, p. 134*)

43 Stems upright   44

44 Leaves pinnately compound with 5–7 sharply toothed leaflets
PACIFIC RED ELDER (*Sambucus racemosa, p. 46*

44 Leaves simple   45

45 Leaves slightly 3-lobed above middle, with sharply toothed margin
SQUASHBERRY (*Viburnum edule, p. 48*)

45 Leaves entire or slightly toothed   46

46 Twigs not angled; leaves blunt-pointed; corolla pink to white; berrylike drupes white
SNOWBERRY (*Symphoricarpos albus, p. 70*)
46 Twigs 4-angled when young: leaves shortpointed; corolla yellow, sometimes tinged with red; berries black   BEARBERRY HONEYSUCKLE (*Lonicera involucrata, p. 68*)

WINTER KEY TO ALASKA SHRUBS

To find the name of a shrub in its winter condition, you must be somewhat of a detective. The first twig seen may not run down quickly in the key. Every bit of evidence will help in finally determining the name. In the field, look for old leaves and remains of flowers and fruits, also well-formed buds of foliage and flowers for the next year. Take notes on the size and general characteristics and look around the area carefully to see whether the specimen is typical.

Becoming familiar with the characters used in the key may save time. Thus, knowing that all shrubs in Alaska with winter buds covered by a single bud-scale are willows (Salix) will aid in learning this genus. Recognition of willows will eliminate running each willow through several steps in the key. When you finally reach a name in the key, check it with the description and range on the map. If either does not agree with your plant, go back through the key to see if alternatives might have been taken along the way. As it is only for native species, this key may not work for shrubs planted around homes.

This winter key to Alaska shrubs is to species except in the willows. Species of willow are not readily distinguished in winter; however, if old dead leaves are present, the vegetative key to willows may be used for further identification.

1   Plants evergreen or with leaves persistent in winter   2
1   Plants deciduous, leafless in winter, dead leaves sometimes persistent   32

Leaves evergreen

2   Plants parasitic on conifer twigs; leaves reduced to paired brownish scales
HEMLOCK DWARF-MISTLETOE (*Arceuthobium tsugense, p. 226*)
2   Plants growing on land; leaves green   3
3   Low shrubs usually less than 1 foot (30 cm) high, mostly forming mats or clumps   4
3   Shrubs usually more than 1 foot (30 cm) high (less in tundra), not forming mats   27
4   Leaves scalelike, awl-shaped or needlelike, narrow   5
4   Leaves relatively broader   15
5   Plants resinous, with persistent berrylike resinous cones, coniferous   6
5   Plants not resinous, not producing berrylike cones, heatherlike   7
6   Leaves awl-shaped, sharp-pointed, spreading in groups of 3
COMMON JUNIPER (*Juniperus communis, p. 19*)
6   Leaves mostly scalelike, blunt, pressed against twig, paired
CREEPING JUNIPER (*Juniperus horizontalis, p. 20*)
7   Leaves less than 1/4 in. (6 mm) long, scalelike or needlelike; twigs without peglike leaf-scars   8
7   Leaves more than 1/4 in. (6 mm) long, needlelike   12
8   Leaves alternate or whorled, spreading, linear or linear-lanceolate, not scalelike   9
8   Leaves paired or opposite, scalelike, pressed against twig   10
9   Leaves mostly 4 in a whorl, sometimes alternate, linear, 1/8–1/4 in. (3–6 mm) long, rounded at tip, with groove on lower surface, hairless; black berries sometimes persistent   CROWBERRY (*Empetrum nigrum, p. 94*)
9   Leaves alternate, linear-lanceolate, 1/16–3/16 in. (2–5 mm) long, pointed, without groove on lower surface, often with long hairs on margin; dried capsule often persistent at tip of twig   STARRY CASSIOPE (*Cassiope stelleriana, p. 87*)

10 Leaves deeply grooved on lower surface, 1/8–3/16 in. (3–5 mm) long
FOUR-ANGLED CASSIOPE (*Cassiope tetragona, p. 88*)
10 Leaves not deeply grooved on lower surface, 1/16–5/32 in. (1.5–4 mm) long   11
11 Twigs with leaves about 1/16 in. (1.5 mm) in diameter
ALASKA CASSIOPE (*Cassiope lycopodioides, p. 85*)
11 Twigs with leaves 1/8 in. (3 mm) in diameter
MERTENS CASSIOPE (*Cassiope mertensiana, p. 86*)
12 Twigs smooth; leaves tightly rolled under, with dense brownish hairs beneath
NARROW-LEAF LABRADOR-TEA (*Rhododendron tomentosum, p. 113*)
12 Twigs with peglike leaf-scars; leaves flat MOUNTAIN-HEATH (*Phyllodoce,* the 3 species below
are not readily distinguished by leaves)   13
13 Leaves short, 3/16–5/16 in. (5–8 mm) long and about 1/16 in. (1.4–1.8 mm) wide
BLUE MOUNTAIN-HEATH (*Phyllodoce coerulea, p. 106*)
13 Leaves longer, more than 5/16 in. (8 mm) long, and narrower, about 1/32 in. (1–1.2 mm)
wide   14
14 Lower surface of leaves with white hairs
ALEUTIAN MOUNTAIN-HEATH (*Phyllodoce aleutica, p. 104*)
14 Lower surface of leaves with reddish resin glands or hairless
RED MOUNTAIN-HEATH (*Phyllodoce empetriformis, p. 108*)
15 Leaves with brown resin dots on both surfaces
LAPLAND ROSEBAY (*Rhododendron lapponicum, p. 112*)
15 Leaves without brown resin glands   16
16 Leaves mostly less than 3/8 in. (10 mm) long   17
16 Leaves more than 3/8 in. (10 mm) long   19
17 Leaves crowded, spatula-shaped, appearing as a whorl at tip of stem
DIAPENSIA (*Diapensia lapponica, p. 74*)
17 Leaves scattered, rounded or linear   18
18 Leaves elliptic, rounded at tip, lower surface with dense hairs; stems coarse, much
branched; plants of dry alpine and arctic tundra
ALPINE-AZALEA (*Kalmia procumbens, p. 100*)
18 Leaves oval to lance-shaped, short-pointed, lower surface hairless; stems very fine, creeping in peatmoss; plants usually in bogs   BOG CRANBERRY (*Vaccinium oxycoccos, p. 120*)
19 Leaves oblong, leathery, with wavy-toothed or straight edges, densely white hairy beneath,
with 2 narrow long-pointed stipules   20
19 Leaves oval or spatula-shaped, not densely white hairy beneath, with stipules   22
20 Leaves wedge-shaped at base; plants mainly pioneers on gravel and sand
DRUMMOND MOUNTAIN-AVENS (*Dryas drummondii, p. 143*)
20 Leaves notched (heart-shaped) at base; plants of alpine tundra or open spruce and shrubs
near tree-line   21
21 Leaves with straight or slightly wavy edges, not or slightly rough above, without glands
and scales on midvein beneath ENTIRE-LEAF MOUNTAIN-AVENS (*Dryas integrifolia, p. 144*)
21 Leaves with wavy-toothed edges, very rough above, with glands and scales on midvein beneath   WHITE MOUNTAIN-AVENS (*Dryas octopetala, p. 148*)
22 Leaves spatula-shaped, broadest at tip and tapering toward base   23
22 Leaves oval, broadest at middle   25
23 Leaves whorled, edges with sharp teeth   PIPSISSEWA (*Chimaphila umbellata, p. 92*)
23 Leaves alternate, edges not toothed   24
24 Leaves without petiole, with conspicuous stiff hairs on edges and lower surface; upright
shrub of alpine tundra of southwest Alaska
KAMCHATKA RHODODENDRON (*Therorhodion camtschaticum, p. 114*)

24 Leaves with petiole 1/8 in. (3 mm) long, hairless on lower surface; reddish berries often persistent; creeping shrub, usually of dry forested area
BEARBERRY (*Arctostaphylos uva-ursi, p. 81*)

25 Leaves not toothed, with edges slightly rolled under
MOUNTAIN-CRANBERRY (*Vaccinium vitis-idaea, p. 124*)

25 Leaves toothed, flat   26

26 Leaves toothed at tip; delicate, creeping, herbaceous shrub
TWINFLOWER (*Linnaea borealis, p. 134*)

26 Leaves finely wavy-toothed; dwarf shrub of Kiska Island in eastern Aleutians
MIQUEL WINTERGREEN (*Gaultheria pyroloides, p. 96*)

27 Leaves 2-4 in. (5-10 cm) long, broad, shiny, sharply toothed on edges
SALAL (*Gaultheria shallon, p. 97*)

27 Leaves less than 2 in. (5 cm) long, narrow, dull, not toothed on edges   28

28 Leaves with dense brownish red curly hairs beneath   29

28 Leaves hairless or nearly so beneath   30

29 Leaves oblong, 1–2 in. (2.5–5 cm) long, 3/16–1/2 in. (5–12 mm) wide, curled downward slightly on edges; fruit stalk bent or curved throughout its length
LABRADOR-TEA (*Rhododendron groenlandicum, p. 110*)

29 Leaves linear, 5/16–5/8 in. (8–15 mm) long, 1/64–1/8 in. (0.5–3 mm) wide, tightly rolled under, curled edges covering lower surface; fruit stalk abruptly bent near capsule
NARROW-LEAF LABRADOR-TEA (*Rhododendron tomentosum, p. 113*)

30 Leaves flat or only slightly rolled under, with scurfy scales often appearing as white dots; young twigs with fine short white hairs   LEATHERLEAF (*Chamaedaphne calyculata, p. 90*)

30 Leaves rolled under, without scurfy scales; twigs hairless   31

31 Leaves elliptic, 1/8–1/2 in. (3–12 mm) wide, slightly rolled under, whitish beneath, veins inconspicuous; southeast Alaska   BOG KALMIA (*Kalmia polifolia, p. 99*)

31 Leaves slightly narrower, 1/16–1/4 in. (2–6 mm) wide, tightly rolled under and partly concealing greenish or whitish lower surface with conspicuous veins; throughout Alaska
BOG ROSEMARY (*Andromeda polifolia, p. 79*)

Leaves deciduous

32 Leaves (or leaf-scars) and twigs opposite or paired   33

32 Leaves (or leaf-scars) and twigs alternate or single   39

33 Twigs and buds covered with minute brown shield-shaped scales
BUFFALOBERRY (*Shepherdia canadensis, p. 77*)

33 Twigs not scaly   34

34 Buds large, more than 3/8 in. (10 mm) long and nearly as broad, stalked; twig stout, dying back at tip; pith broad   PACIFIC RED ELDER (*Sambucus racemosa, p. 46*)

34 Buds small, mostly less than 3/8 in. (10 mm) long, if longer then less than 3/16 in. (5 mm) wide, stalked or stalkless; twig slender, usually not dying back; pith narrow   35

35 Twigs 4-angled or squarrish   BEARBERRY HONEYSUCKLE (*Lonicera involucrata, p. 68*)

35 Twigs round   36

36 Leaf-scars raised, often torn or indistinct, with 1 bundlescar; twigs very slender with bark becoming shreddy   SNOWBERRY (*Symphoricarpos albus, p. 70*)

36 Leaf-scars not raised or torn, with 3 or more bundle-scars; twigs less slender, with bark not shreddy   37

37 Twigs red, shiny, densely gray hairy near tip; buds gray brown
RED-OSIER DOGWOOD (*Cornus alba, p. 71*)

37 Twigs gray, hairless, buds red   38

38 Buds long, narrow, pointed, dark brownish red, outer bud-scales united at edges, inner bud-scales hairless; loose straggling shrubs   SQUASHBERRY (*Viburnum edule, p. 48*)

38 Buds rounded, blunt, bright red, outer bud-scales often spreading and exposing hairy inner bud-scales; erect shrub or small tree with maple key fruits often persistent
DOUGLAS MAPLE (*Acer glabrum, p. 227*)

39 Twigs with spines, thorns; or prickles (often absent on young plants and new shoots, especially in *Rubus spectabilis*) 40

39 Twigs without spines, thorns, or prickles 48

40 Twigs very stout, light brown, densely covered with slender sharp spines
DEVIL'S-CLUB (*Oplopanax horridus, p. 50*)

40 Twigs slender, of various colors, with spines less dense or partly enlarged at base 41

41 Spines stout, 3/8–1 in. (10–25 mm) long, few on shiny red brown twigs; purplish black berries often persistent BLACK HAWTHORN (*Crataegus douglasii, p. 140*)

41 Spines less than 1/2 in. (12 mm) long 42

42 Spines 3–9 at nodes and smaller single spines between; pith with spongelike cavities; shrubs usually trailing SWAMP GOOSEBERRY (*Ribes lacustre, p. 130*)

42 Spines single; pith not spongelike; erect shrubs 43

43 Twigs light brown or whitish, soft and easily broken, bark usually shreddy, pith occupying more than 2/3; old raspberries often present RASPBERRY (*Rubus*) 44

43 Twigs dark red, hard, bark not shreddy, pith occupying less than 2/3; old rose hips often present rose (*Rosa*) 46

44 Twigs whitish, with stout hooked flattened prickles or spines
WESTERN BLACK RASPBERRY (*Rubus leucodermis, p. 159*)

44 Twigs brown 45

45 Twigs straight, covered with bristles and pricklesAmerican red raspberry (*Rubus idaeus, p. 158*)

45 Twigs zigzag, with weak straight rounded pricklesSALMONBERRY (*Rubus spectabilis, p. 162*)

46 Twigs with few flattened prickles or spines usually paired at nodes
NOOKTA ROSE (*Rosa nutkana, p. 156*)

46 Twigs with prickles or spines round or partly so, many to few 47

47 Prickles or spines many PRICKLY ROSE (*Rosa acicularis, p. 154*)

47 Prickles or spines few, scattered WOODS' ROSE (*Rosa woodsii, p. 157*)

48 Shrubs low, less than 6 in. (15 cm) high or dying back to woody base 49

48 Shrubs usually upright and more than 6 in. (15 cm) high 54

49 Shrubs creeping 50

49 Shrubs herbaceous, dying back to woody base; dead leaves often persistent, divided into narrow segments 52

50 Buds covered by a single bud-scale DWARF WILLOWS (*Salix* spp.; if old dead leaves are present, the vegetative key to willows may be used for identification to species)

50 Buds with 2 or more bud-scales 51

51 Red or brown leaves or skeletonized leaves persisting, black berries often persistent
ALPINE BEARBERRY (*Arctous alpina, p. 82*)

51 Leaves shedding first year; red berries often persistent
RED-FRUIT BEARBERRY (*Arctous rubra, p. 56*)

52 Leaves twice 3-forked, hairless, without odor LUETKEA (*Luetkea pectinata, p. 150*)

52 Leaves finely dissected, whitish hairy, with sagebrush odor persisting 53

53 Basal leaves 1/4–1/2 in. (6–12 mm) long, 2–3 times divided into narrow segments 1/32 in. (1 mm) wide FRINGED SAGEBRUSH (*Artemisia frigida, p. 53*)

53 Basal leaves 1–2 in. (2.5–5 cm) long, 2–3 times divided into spatula-shaped segments 1/16–1/8 in. (2–3 mm) wide ALASKA SAGEBRUSH (*Artemisia alaskana, p. 52*)

54 Twigs with expanded buds of next year's catkins, remains of last year's catkins, and conspicuous dots (lenticels or resin glands) **55**

54 Twigs without catkins **60**

55 Twigs resinous, buds of next year's catkins small, 1/4 in. (6 mm) long and stalkless, covered by several whitebordered bud-scales; remains of last year's catkin spikelike; winter buds not stalked, of overlapping bud-scales **56**

55 Twigs not resinous, buds of next year's catkins 3/8–5/8 in. (10–15 mm) long on stalks of 3/16-3/8 in. (5-10 mm), bud-scales not white-bordered; old, hard blackish cones or cone-like fruits usually present ALDER (*Alnus*) **58**

56 Remains of last year's catkin a stalkless straight stout spike 3/8 in. (10 mm) long, 1/16 in. (1.5 mm) wide, with conspicuous concave bud-scars; resin dots inconspicuous, on young twigs only SWEETGALE (*Myrica gale, p. 136*)

56 Remains of last year's catkin very narrow, 3/8–5/8 in. (10–15 mm) long, 1/64 in. (0.5 mm) wide, long stalked; twigs covered with resin glands BIRCH (*Betula*) **57**

57 Shrubs usually less than 2 feet (0.6 m) high, in bogs and tundra; catkin scale without resiniferous dot or hump, glandless; broad wing around nutlet
DWARF ARCTIC BIRCH (*Betula nana, p. 64*)

57 Shrub to 5 feet (1.5 m) high, near tree line; catkin scale with resinous dot or hump, often glandular; wing of nutlet narrow, often broader toward tip
RESIN BIRCH (*Betula glandulosa, p. 60*)

58 Winter buds with 3 exposed scales meeting at edges SITKA ALDER (*Alnus viridis, p. 58*)

58 Winter buds of overlapping scales **59**

59 Cones 1/2–1 in. (12–25 mm) long RED ALDER (*Alnus rubra, p. 56*)

59 Cones less than 1/2 in. (12 mm) long THINLEAF ALDER (*Alnus incana, p. 55*)

60 Stipules and bases or stumps of petioles persistent, partly covering buds **61**

60 Stipules and bases of petioles absent **62**

61 Stipules narrow, bent or twisted; twigs soft, canelike, dying back from tip
WESTERN THIMBLEBERRY (*Rubus parviflorus, p. 160*)

61 Stipules broad, papery, spreading; twigs hard, not dying back
BUSH CINQUEFOIL (*Dasiphora fruticosa, p. 141*)

62 Fruits persistent in conspicuous clusters **63**

62 Fruits absent or borne singly **69**

63 Fruits fleshy, like a small apple, red; winter buds large, mostly more than 3/8 in. (10 mm) long, with densely hairy inner bud-scales; large shrubs and small trees MOUNTAIN-ASH (*Sorbus*) **64**

63 Fruits dry, 3–5 from a flower, egg-shaped, podlike, splitting open, brown; winter buds less than 3/8 in. (10 mm) long **67**

64 Winter buds with whitish hairs **65**

64 Winter buds with rusty brown hairs **66**

65 Winter buds reddish brown, inner bud-scales with whitish hairs at tip
CASCADE MOUNTAIN-ASH (*Sorbus scopulina, p. 167*)

65 Winter buds densely covered with whitish hairs; naturalized tree
EUROPEAN MOUNTAIN-ASH (*Sorbus aucuparia, p. 164*)

66 Winter buds dull reddish brown, densely rusty hairy
SITKA MOUNTAIN-ASH (*Sorbus sitchensis, p. 168*)

66 Winter buds shiny reddish brown, slightly rusty hairy; only in westernmost Aleutian Islands SIBERIAN MOUNTAIN-ASH (*Sorbus sambucifolia, p. 166*)

67 Fruits 1/4–3/8 in. (6-10 mm) long; bark peeling and shedding in long strips
PACIFIC NINEBARK (*Physocarpus capitatus, p. 152*)

67 Fruits less than 1/8 in. (3 mm) long; bark not shedding   **68**

68 Fruit clusters flat-topped to half round   BEAUVERD SPIREA (*Spiraea stevenii, p. 172*)

68 Fruit clusters conic, much longer than broadDOUGLAS SPIRAEA (*Spiraea douglasii, p. 170*)

69 Winter buds covered by a single scale willow (*Salix*; species not readily distinguished in winter; old leaves and catkins sometimes can be found for identification to species using the *Vegetative Key to Alaska Willows*. Descriptions, size of plants, and range maps may also be helpful.)   WILLOWS (*Salix, p. 179*)

69 Winter buds with 2 or more scales exposed   **70**

70 Twigs with rusty brown scales when young, becoming silvery; silvery berries often persistent   SILVERBERRY (*Elaeagnus commutata, p. 76*)

70 Twigs and fruits not as above   **71**

71 Twigs without end buds; side buds with 2 budscales meeting at edges (except in *Vaccinium uliginosum*); fruit a blue or red berry, seldom persistent BLUEBERRIES AND HUCKLE-BERRIES (*Vaccinium*)   **72**

71 Twigs with true end buds covered by 3 or more bud-scales   **76**

72 Shrubs mostly less than 16 in. (40 cm) high; twigs round or sometimes slightly angled   **73**

72 Shrubs mostly more than 2 ft. (60 cm) high; twigs angled   **74**

73 Bud-scales 2, meeting at edges   DWARF BLUEBERRY (*Vaccinium cespitosum, p. 118*)

73 Bud-scales several, overlapping   BOG BLUEBERRY (*Vaccinium uliginosum, p. 123*)

74 Twigs green, strongly angled; fruit red   RED HUCKLEBERRY (*Vaccinium parvifolium, p. 122*)

74 Twigs brown or reddish, weakly angled; fruit blue or black   **75**

75 Fruit stalks usually less than 3/8 in. (1 cm) long, curved, not enlarged below fruit   EARLY BLUEBERRY (*Vaccinium ovalifolium, p. 119*)

75 Fruit stalks often more than 3/8 in. (1 cm) long, straight or nearly so, enlarged just below fruit   ALASKA BLUEBERRY (*Vaccinium alaskaense, p. 116*)

76 Shrubs spreading; twigs angled, with papery shedding or shredded bark, often with unpleasant odor when crushed; pits porous or spongy CURRANTS (*Ribes*)   **77**

76 Shrubs erect (or becoming small trees); twigs rounded, with bark not shedding (except in *Elliottia pyroliflora*); pith hard, solid   **81**

77 Twigs stout, 1/4 in. (6 mm) in diameter; leafscars heart-shaped, large, gray   STINK CURRANT (*Ribes bracteosum, p. 126*)

77 Twigs slender, less than 3/16 in. (5 mm) in diameter; leaf-scars V-shaped, narrow and inconspicuous   **78**

78 Twigs with black gland dots   **79**

78 Twigs without gland dots   **80**

79 Buds hairless; twigs 1/16 in. (2 mm) in diameter   NORTHERN BLACK CURRANT (*Ribes hudsonianum, p. 129*)

79 Buds with white hairs; twigs about 1/8 in. (3 mm) in diameter   SKUNK CURRANT (*Ribes glandulosum, p. 128*)

80 Twigs hairy, yellow brown, becoming dark brown, about 3/16 in. (5 mm) in diameter trailing   BLACK CURRANT (*Ribes laxiflorum, p. 131*)

80 Twigs hairless, light brown, becoming reddish brown and shredded, about 1/8 in. (3 mm) in diameter   AMERICAN RED CURRANT (*Ribes triste, p. 133*)

81 Twigs paired or whorled, widely forking, with gland hairs, odorous when crushed   RUSTY MENZIESIA (*Menziesia ferruginea, p. 102*)

81 Twigs not paired, without gland hairs   **82**

82 Winter buds orange   COPPERBUSH (*Elliottia pyroliflora, p. 93*)

82 Winter buds darker   **83**

83 Winter buds blunt-pointed, dark brown; twigs coarse, gray or brown, often with dense gray hairs near tip, with short side twigs or spurs

OREGON CRAB APPLE (*Malus fusca, p. 151*)

83 Winter buds sharp-pointed, purple; twigs slender, reddish purple, shiny, hairless, without short side twigs or spurs   **SASKATOON SERVICEBERRY** (*Amelanchier alnifolia, p. 138*)

COPPERBUSH, *p. 93*

GLOSSARY

ACHENE small, dry and hard one-seeded fruit.

ALTERNATE LEAVES leaves arranged on alternating sides of the twig.

ANGIOSPERM class of plants that has the seeds enclosed in an ovary; includes flowering plants.

AWL-LIKE LEAVES short leaves that taper evenly to a point; found on junipers and redcedars.

BERRY fleshy fruit with several seeds.

BISEXUAL FLOWER a perfect flower; a flower with organs of both sexes present.

BROADLEAF trees having broad, flat-bladed leaves rather than needles; also a common name for hardwoods.

CAMBIUM layer of tissue one to several cells thick found between the bark and the wood; divides to form new wood and bark.

CAPSULE dry fruit that splits open, usually along several lines, to reveal many seeds inside.

CATKIN a cylindrical flower cluster, with inconspicuous or no petals (as in willows, *Salix*), usually wind-pollinated but sometimes insect-pollinated. They contain many, usually unisexual flowers, arranged closely along a central stem. Also called an *ament.*

CHAMBERED PITH pith divided into many empty horizontal chambers by cross partitions.

COMPOUND LEAVES leaves with more than one leaflet attached to a stalk called a rachis.

CONIFER trees and shrubs that usually bear their seeds in cones and are mostly evergreen; includes cypresses, pines, firs, spruces, and yews.

DECIDUOUS LEAVES leaves that die and fall off trees after one growing season.

DICHOTOMOUS KEY a key to tree identification based on a series of decisions, each involving a choice between two alternate identification characteristics.

DIOECIOUS having unisexual flowers with staminate (male) and pistillate (female) flowers borne on different trees.

DRUPE fleshy fruit with a single stone or pit.

ELLIPTIC resembling an ellipse and about one-half as wide as long.

ENTIRE MARGIN leave margins that are smooth (not toothed).

EVERGREEN trees and shrubs that retain their live, green leaves during the winter and for two or more growing seasons.

FAMILY group of closely related species and genera; scientific name ends in "aceae".

GENUS a group of species that are similar; the plural of genus is genera.

GLABROUS Smooth, with no hair or scales.

GYMNOSPERM large class of plants having seeds without an ovary, usually on scales of a cone; includes conifers and the ginkgo.

HARDWOODS usually refers to trees that have broad-leaves and wood made up of vessels; similar to angiosperms.

HEARTWOOD nonliving wood (often dark) found in the middle of a tree's stem.

IMPERFECT flower a unisexual flower with either functional stamens or pistils but not both.

INFLORESCENCE the flowering portion of a plant.

LANCEOLATE lance-shaped; about 4 times as long as wide and widest below the middle.

LATERAL BUDS buds found along the length of the twig (not at the tip); they occur where the previous year's leaves were attached.

LEAFLETS small blades of a compound leaf attached to a stalk (rachis); without buds where they attach.

LOBED MARGIN leaf margin with gaps that extend more or less to the center of the leaf.

LUSTROUS glossy, shiny.

MONOECIOUS having unisexual flowers with staminate (male) and pistillate (female) flowers borne on the same tree, though often on different branches.

NATURALIZED nonnative trees that have escaped cultivation and are growing in the wild.

NEEDLE-LIKE LEAVES very thin, sharp, pointed, pin-like leaves; found on pines, firs and some other softwoods.

NODE the point on a stem at which leaves and buds are attached.

OBOVATE inversely ovate.

OPPOSITE LEAVES leaves arranged directly across from each other on the twig.

ORBICULAR circular in outline.

OVAL broadly elliptic, with the width greater than one-half the length.

OVATE having the lengthwise outline of an egg, widest below the middle.

PALMATELY COMPOUND compound leaves in which several leaflets radiate from the end of a stalk (rachis); like the fingers around the palm of a hand.

PERFECT FLOWER a bisexual flower with functional stamens and pistils.

PERSISTENT LEAVES leaves that remain on the tree during winter.

PETIOLE a slender stalk that supports a simple leaf.

PHOTOSYNTHESIS process through which the leaves, with energy from sunlight, make food from water and carbon dioxide.

PINNATELY COMPOUND compound leaves in which leaflets are attached laterally along the rachis or stalk; leaves may be once, twice, or three-times pinnately compound.

PISTIL the ovary-bearing (female) organ of a flower.

PISTILLATE FLOWER a unisexual (female) flower bearing only pistils.

PITH soft and spongy, or chambered tissue found in the middle of the stem.

POLYGAMO-DIOECIOUS having unisexual flowers with staminate (male) and pistillate (female) flowers borne on different trees, but also having some perfect flowers on each tree.

POLYGAMO-MONOECIOUS having unisexual flowers with staminate (male) and pistillate (female) flowers borne on the same tree, along with some perfect flowers on each tree.

POLYGAMOUS Having some unisexual flowers and some bisexual flowers on each plant (can be polygamo-monoecious or polygamo-dioecious).

POME fruit with a fleshy outer coat and a stony layer (similar to plastic) within, with seeds inside the stony layer (apples, pears, etc.).

PUBESCENT covered with hairs.

RACHIS the central stalk to which leaflets of a compound leaf are attached.

RHOMBIC with an outline resembling a rhombus (diamond-shaped).

SAMARA dry fruit with one or two flat wings attached to a seed (as on elms and maples).

SAPWOOD living wood, often light colored, found between the bark or cambium and the heartwood, usually darker colored.

SCALE-LIKE LEAVES small, short, fish-scale-like leaves which cover the entire twig; found on juniper and redcedar.

SCIENTIFIC NAMES Latin-based names used world-wide to standardize names of trees and other plants and animals.

SERRATE with teeth.

SHADE INTOLERANT trees that need a lot of sunlight for growth and survival.

SHADE TOLERANT trees that can tolerate less sunlight for growth and survival.

SHRUB low-growing woody plant with many stems rather than one trunk.

SIMPLE LEAVES leaves with one blade attached to a petiole, or stalk.

SINUS a recess between two leaf lobes.

SOFTWOODS usually refers to trees that are conifers or cone-bearing; conifers generally have softer wood than angiosperms or hardwoods, but there are many exceptions.

SOLID PITH pith that is not divided into chambers.

SPECIES trees with similar characteristics and that are closely related to each other; 'species' is used in both the singular and plural sense.

SPRING-WOOD wood on the inside of an annual ring, formed during the spring; cells are often thinner-walled.

STAMEN the pollen-bearing (male) organ of a flower.

STAMINATE FLOWER a unisexual (male) flower bearing only stamens.

STROBILE a cone or inflorescence with over-lapping bracts or scales.

SUMMER-WOOD wood on the outside of an annual ring, formed during the summer; this wood is sometimes dark and cells are often thicker-walled.

TEPAL A usually showy part of the outer portion of a flower that is not differentiated into a sepal or petal.

TERMINAL BUDS bud appearing at the tip, or end, of a twig; usually larger than other lateral buds.

TOOTHED/SERRATED MARGIN leaf margin with coarse, fine, sharp or blunt teeth.

TRACHEIDS small-diameter tubes in the wood of trees that carry water from the roots to the leaves; water carrying tubes in conifer xylem are all tracheids.

TREE a woody plant with one to a few main stems and many branches; usually over 10 feet (3 m) tall.

UNISEXUAL FLOWER an imperfect flower; a flower with organs of only one sex present.

XYLEM the wood of a tree, made up of strong fibers, tracheids and vessels.

NETLEAF WILLOW, p. 216

ACKNOWLEDGMENTS

The basis of this book was the classic *Alaska Trees and Shrubs,* by Leslie A. Viereck and Elbert L. Little, Jr., published in 1972 by the USDA Forest Service as Agriculture Handbook 410. After nearly 50 years, much has changed in terms of species nomenclature, and our knowledge of each species' distribution and habitat patterns. distribution information. The current work updates the taxonomic nomenclature (i.e., the scientific plant names), provides current range distribution maps (based on herbarium specimens), adds color photographs, and extensively revises the original text. The authors of the Forest Service book were highly respected experts in their fields, and a brief biography of each follows.

Leslie A. Viereck first studied the forests and vegetation of Alaska during the summer of 1948 while he was still an undergraduate at Dartmouth College. He completed his graduate work (M.S. and Ph.D.) at the University of Colorado under the direction of William A. Weber, the pre-eminent Colorado botanist, and John W. Marr, the founder and first Director of the Institute of Arctic and Alpine Research. For his doctoral thesis he conducted a study of plant succession and soil development on gravel outwash of the Muldrow Glacier, Alaska. In 1959 Viereck permanently moved to Alaska as Assistant Professor of Botany at the University of Alaska in Fairbanks, where he continued his long-time research of the Alaskan boreal forest. Viereck passed away at age 78 in August 2008 in Fairbanks, Alaska.

Elbert Luther Little, Jr., was a botanist and who specialized in the study of trees. He was born in Fort Smith, Arkansas on October 15, 1907. He earned a B.A. in Botany from the University of Oklahoma (1927) a M.S. and Ph.D. in botany at the University of Chicago (1929), and a B.S. in zoology by the University of Oklahoma (1932). Little worked as a forest ecologist with the USDA Forest Service in the southwest United States, focusing on poisonous range plants and pinyon-juniper woodlands. He eventually worked as a dendrologist with the U.S. Forest Service for 34 years, becoming chief dendrologist in 1967. He retired in 1975. In addition to Alaska, during his career Little spent time in Central and South America collecting plants, working in conjunction with the United Nations, and as a visiting professor at universities in Venezuela and Costa Rica. Little passed away in 2004.

Range maps were prepared from data provided by the Consortium of Pacific Northwest Herbaria and the E-Flora BC (Electronic Flora of British Columbia); see Resources for website addresses. Photographs were obtained, where possible, from public domain sources, from the author's own collection, and from a number of photographers on Flickr who have made their work available under Creative Commons commercial-use licences (see www.flickr.com). Detailed plant illustrations were generated from herbarium specimens housed at a number of herbaria in the United States and Canada.

Plant nomenclature generally follows the accepted names as listed in the *Integrated Taxonomic Information System* (ITIS; http://www.itis.gov). Additional taxonomic and distributional data were provided by the *Biota of North America Program* (BONAP; www.bonap.org)

RESOURCES

Print and online resources consulted in the preparation of this book are listed below.

Print

Anderson, J.P. 1959. *Flora of Alaska and Adjacent Parts of Canada. An illustrated descriptive text of all vascular plants known to occur within the region covered.* Iowa State University Press. Ames, Iowa.

Argus, G.W. 1973. *The Genus Salix in Alaska and the Yukon.* Publications in Botany No. 2. National Museum of Canada. Ottawa, Ontario.

Argus, G.W. 2004. *A Guide to the identification of Salix (willows) in Alaska, the Yukon Territory, and adjacent regions.* July 2004 workshop on willow identification. Unpublished.

Cody, W.J. 2000. *Flora of the Yukon Territory.* Second Edition. National Research Council of Canada Press. Ottawa, Ontario.

Collet, D. 2004. *Willows of Interior Alaska.* U.S. Fish and Wildlife Service.

Douglas, G.W., G.B. Straley, D.V. Meidinger, and J. Pojar. 1998. *Illustrated Flora of British Columbia.* Volumes 1-8. B.C. Ministry of Environment, Land, & Parks and B.C. Ministry of Forests. Victoria, British Columbia.

Flora of North America Editorial Committee (eds.). 1993+. *Flora of North America North of Mexico.* 20+ vols. New York & Oxford.

Hitchcock, C.L., A. Cronquist, D.E. Giblin, B.S. Legler, P.F. Zika, and R.G. Olmstead. 2018. *Flora of the Pacific Northwest: An Illustrated Manual.* Second Edition. University of Washington Press. Seattle, Washington.

Hitchcock, C.L., A. Cronquist, M. Ownbey, and J.W. Thompson. 1955-1969. *Vascular Plants of the Pacific Northwest.* University of Washington Press. Seattle, Washington.

Hultén, E.O. 1968. *Flora of Alaska and Neighboring Territories: A Manual of the Vascular Plants.* Stanford University Press. Stanford, California.

Johnson, D., L.J. Kershaw, A. MacKinnon, and J. Pojar. 1995. *Plants of the Western Boreal Forest and Aspen Parkland.* Lone Pine Publishing. Vancouver, British Columbia.

Kuchler, A.W. 1964. *Potential Natural Vegetation of the Conterminous United States.* American Geographical Society, Special Publication No. 36

Tolmatchev, A.I., J.G. Packer, and G.C.D. Griffiths. 1996. *Flora of the Russian Arctic.* Volumes 1-3. University of Alberta Press. Edmonton, Alberta.

Viereck, L.A., and E.L. Little. 2007. *Alaska Trees and Shrubs.* Second Edition. University of Alaska Press. Fairbanks, Alaska.

Welsh, S.L. 1974. *Anderson's Flora of Alaska and Adjacent Parts of Canada.* Brigham Young University Press. Provo, Utah.

Online

Alaska Center for Conservation Science (ACCS) (https://accs.uaa.alaska.edu/)
Alaska Native Plant Society (AKNPS) (https://aknps.org/
Consortium of Pacific Northwest Herbaria (http://www.pnwherbaria.org/index.php)
E-Flora British Columbia (https://ibis.geog.ubc.ca/biodiversity/eflora/)
Arctic Flora of Canada and Alaska (http://arcticplants.myspecies.info/)
Ecological Atlas of Denali's Flora, Denali National Park and Preserve (https://ecologicalatlas.uaf.edu/)

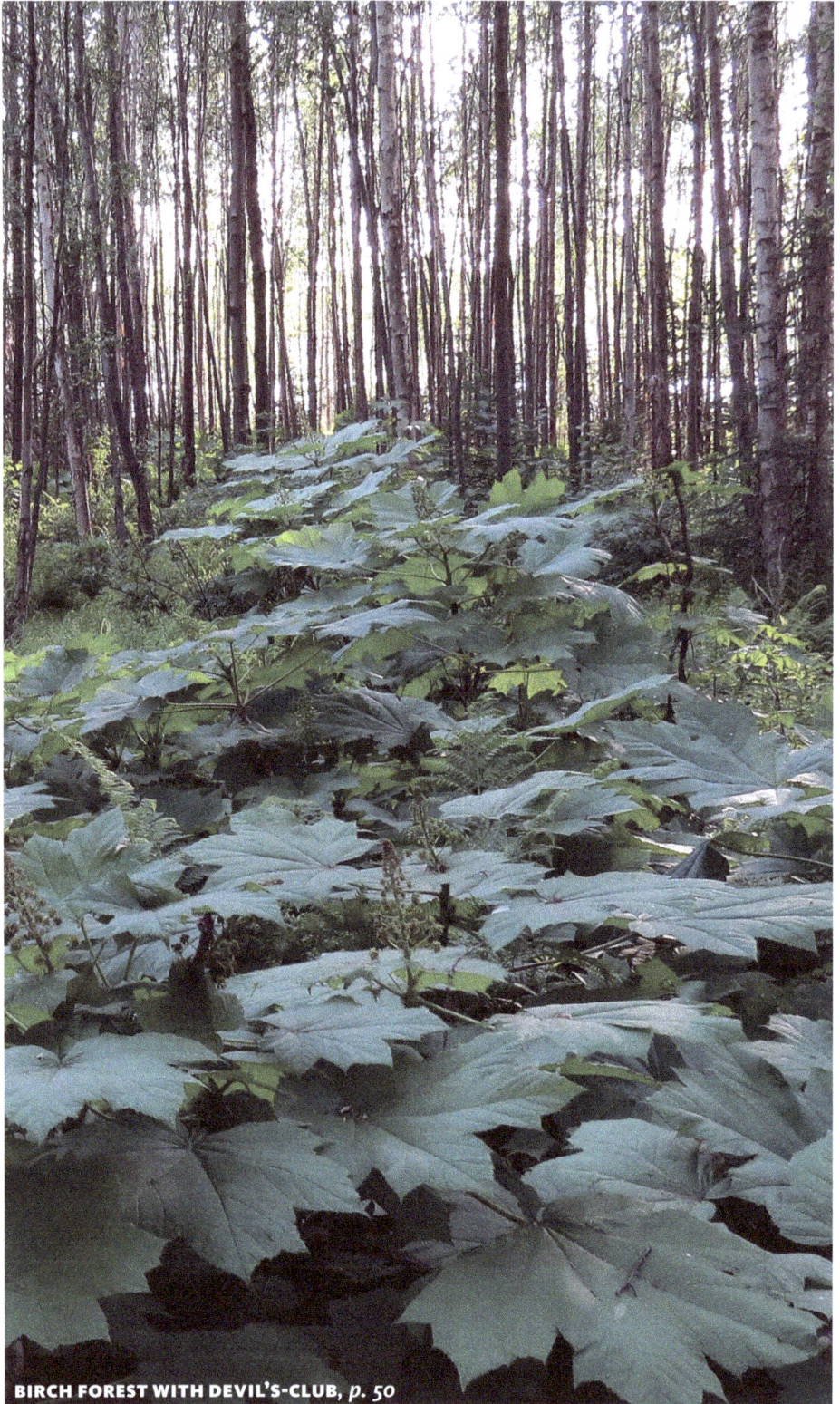

BIRCH FOREST WITH DEVIL'S-CLUB, *p. 50*

INDEX — COMMON NAME

BOG BLUEBERRY, *p. 123*

INDEX — SCIENTIFIC NAME